Research and Management of High-Elevation Five-Needle Pines in Western North America

Proceedings of The Second
High-Five Conference:
October 5-7, 2021

FIG. 11.—*Pinus balfouriana: a,* seeds.

The Press at Cal Poly Humboldt

Cal Poly Humboldt Library
1 Harpst St.
Arcata, CA 95524-8299
digitalcommons.humboldt.edu

ISBN: 978-1-947112-78-0

Layout & design: Alannah Guevara & Aaron Laughlin

Cover Photos & Credits - Clockwise from top:

Foxtail Pine (*Pinus balfouriana*) south of Cottonwood Pass, CA (credit: MP Murray)
Great Basin Bristlecone Pine (*P. longaeva*) on Cave Mountain, NV (credit: C Earle, www.conifers.org)
Whitebark Pine (*P. albicaulis*) at Galena Summit, ID (credit: MP Murray)
Rocky Mountain Bristlecone Pine (*P. aristata*), Mt. Goliath Natural Area, CO (credit: MP Murray)
Southwestern White Pine (*P. strobiformis*) near El Salto, Durango, CO (credit: C Earle, www.conifers.org)
Limber Pine (*P. flexilis*) at Columbia Lake, BC (credit: MP Murray)

Citation

Murray, Michael P.; Smith, Cyndi M.; McKinney, Shawn T.; Achuff, Peter L. 2022. Research and management of high-elevation five-needle pines in western North America: Proceedings of the second High-Five Conference. 5-7 October 2021. The Press at Cal Poly Humboldt

All papers in these proceedings were formatted to style standards with some minor editing. Authors are responsible for content and accuracy of their individual papers and the quality of illustrative materials.

The Press at Cal Poly Humboldt publishes high-quality scholarly, intellectual, and creative works by or in support of our campus community. The Press supports the Cal Poly Humboldt purpose to improve the human condition and our environment by promoting understanding of social, economic, and environmental issues.

Contents

Preface

Taking place from October 5-7, 2021, this second high-five (H5II) conference brought together the world's leading experts on North America's magnificent high-elevation pines. The high-five conferences are for scientists and management professionals interested in learning about new management techniques, research findings, and the latest trends in high-elevation five-needle pine ecosystems. Hosted by the Whitebark Pine Ecosystem Foundation (WPEF), 287 registrants benefitted from 97 presentations plus a dozen keynote and plenary talks. This proceedings document captures submitted papers plus all abstracts of presentations and posters that covered a tremendous variety of topics. Keynote and plenary talks are not included as papers here, as these were submitted as a special issue to will be published in the Journal of Forest Ecology and Management.

Originally planned as a gathering in Missoula, MT during 2020, a viral pandemic (Covid-19) emerged - presenting very unexpected challenges. As a result, the conference presentations were pre-recorded for viewing on-screen. Aired October 5-7, each recording was followed by a live question-answer period. Similarly, field trip presenters provided their field recordings for a very unique viewing experience.

Why is WPEF investing so much time and effort into this conference? Many high-elevation five-needle pine forests are rapidly declining throughout North America. The six high-elevation five-needle pine species — whitebark (*Pinus albicaulis* Engelm.), limber (*P. flexilis* James), southwestern white (*P. strobiformis* Engelm.), foxtail (*P. balfouriana* Grev. & Balf.), Great Basin bristlecone (*P. longaeva* D.K. Bailey), and Rocky Mountain bristlecone (*P. aristata* Engelm.) — are of great ecological and symbolic importance to both the U.S. and Canada. The objectives of our series of high-five conferences are to: (1) bring together scientists, managers, and concerned citizens to exchange information on the ecology and management of these important pines; (2) learn about the threats and current status of pine populations; (3) describe efforts to mitigate threats through restoration techniques and action plans; and (4) build a foundation for the synthesis of research efforts and management approaches.

This conference was made possible by the steering committee, collaborators, and sponsors. We appreciate the generous support provided by American Forests; British Columbia Forest Service; Montana Department of Natural Resources and Conservation; National Forest Foundation; Montana Native Plant Society; Society of American Foresters; Northern Rockies Fire Science Network; College of Forestry & Conservation at the University of Montana; Crater Lake Institute; The Nature Conservancy – Montana; The Nature Conservancy – Idaho; Charles Bacon & Cynthia Dusel-Bacon; Horizon Credit Union, Missoula, MT; Dr. Cathy Cripps, Montana State University; and the John & Janet Tagney Charity.

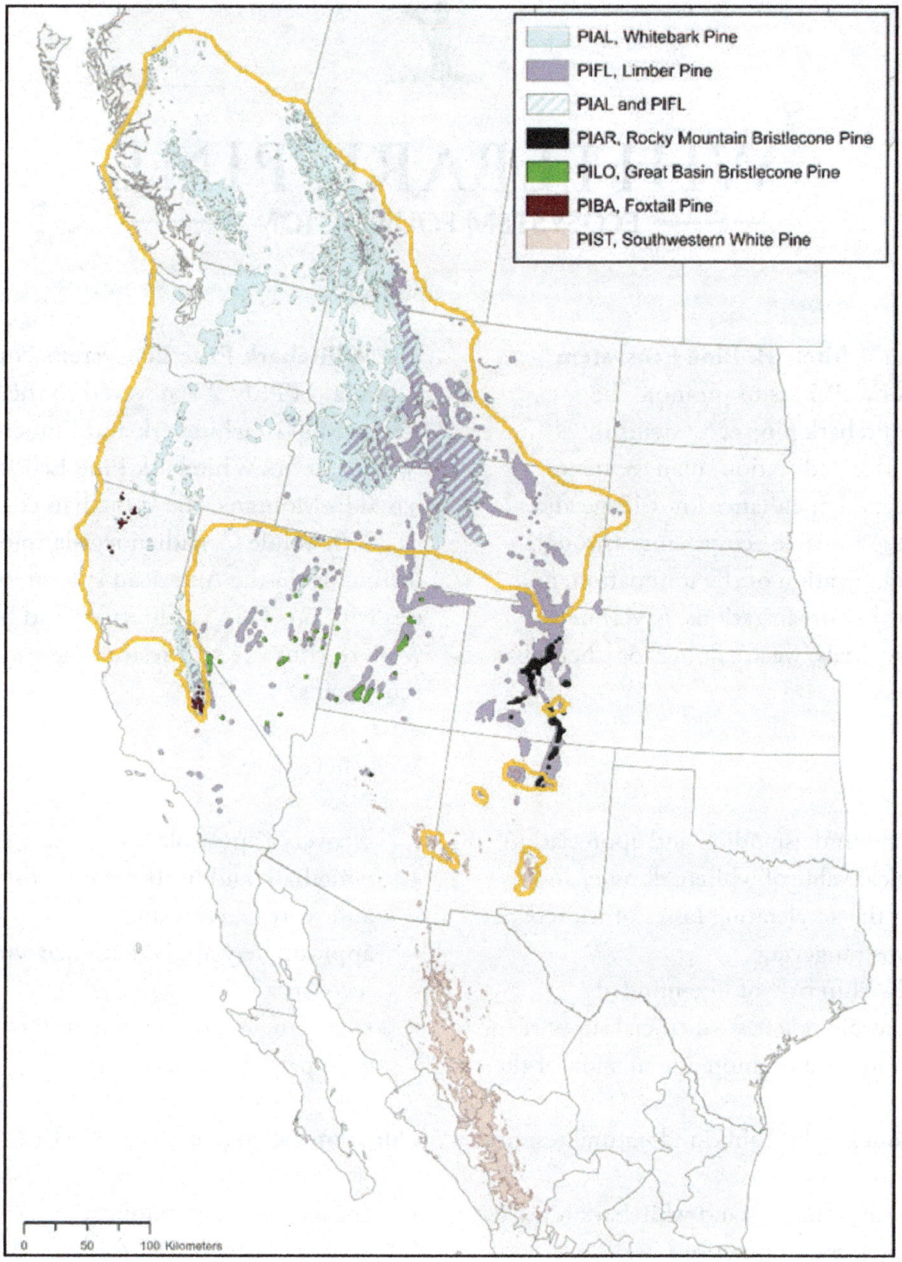

Distribution of the high-elevation five-needle pines and estimated extent of white pine blister rust (orange polygons) in western North America. Base maps were compiled by Schoettle et al. (2022) and updated blister rust distribution adapted from Schoettle et al. (2019).

Schoettle AW, KS Burns, ST McKinney, J Krakowski, KM Waring, DF Tomback, and M Davenport. 2022. Integrating forest health condition and species adaptive capacities to infer and affect future trajectories of the high-elevation five-needle pines. Forest Ecology and Management: in review.

Schoettle AW, KS Burns, CM Cleaver, and JJ Connor. 2019. Proactive limber pine conservation strategy for the Greater Rocky Mountain National Park Area. General Technical Report RMRS-GTR-379. Fort Collins, CO: U.S. Department of Agriculture, Forest Service, Rocky Mountain Research Station. 81p. https://www.fs.usda.gov/treesearch/pubs/57621

WHITEBARK PINE
ECOSYSTEM FOUNDATION

The Mission of the **Whitebark Pine Ecosystem Foundation USA** (WPEF) is to promote the conservation of whitebark pine ecosystems by supporting restoration, education, management, and research projects that enhance knowledge and stewardship of these valuable ecosystems. Through the cooperative organization of the foundation, and networking with other organizations, government agencies, and individuals, we are doing together what we cannot do alone.

We:

1. Promote general understanding and appreciation for the ecological value of whitebark pine, and recognition of the accelerating losses of whitebark pine ecosystems rangewide.
2. Form a membership base of like-minded individuals who provide basic financial support for operational needs and promote the mission of the WPEF.
3. Inform and educate the public and natural resource agencies.
4. Support research to understand whitebark pine ecosystem processes and functions.
5. Test and promote possible strategies for restoration.
6. Provide financial assistance and technical support for restoration efforts.
7. Form associations with other organizations, governmental agencies, and individuals to further cooperation and collaboration in education, research, and restoration projects related to whitebark pine ecosystems.

USA
whitebarkfound.org

The **Whitebark Pine Ecosystem Foundation of Canada** (WPEFC) is devoted to the conservation and stewardship of whitebark and limber pine ecosystems. A sister to the Whitebark Pine Ecosystem Foundation based in Montana, the Canadian chapter was founded in 2010. While Canadian legislation and policies are distinct from the American system, we maintain strong ties with our USA counterpart and both organizations work to conserve and restore these vulnerable ecosystems.

We believe that:

- recovery is possible
- whitebark and limber pine ecosystems are valuable and worth recovering
- appropriately applied science will aid in this recovery
- engagement, connection, and communication with partners is necessary

Our strategic goals are that WPEFC:

- engages with the public and partners using a variety of diverse platforms
- has sufficient funds to deliver key objectives
- is an integral and respected hub for knowledge sharing
- is governed by an engaged and effective Board of Directors
- plays a leading role in coordinating the conservation & recovery of whitebark and limber pine ecosystems across their Canadian Ranges

CANADA
whitebarkpine.ca

THANK YOU TO OUR SPONSORS

Whitebark Pine Level ($5,000+)
American Forests

Limber Pine Level ($1,000-$4,999)
Montana Dept. of Natural Resources and Conservation
National Forest Foundation
Montana Native Plant Society
Society of American Foresters

Bristlecone Pine Level ($600-$999)
Northern Rockies Fire Science Network
W.A. Franke College of Forestry & Conservation (U of MT)

Foxtail Pine Level ($300-$599)
Crater Lake Institute
The Nature Conservancy - Montana
The Nature Conservancy - Idaho
Charles Bacon & Cynthia Dusel-Bacon

Nutcracker Level ($100-$299)
Horizon Credit Union, Missoula, MT
Dr. Cathy Cripps, Montana State University
John & Janet Tagney Charity

Conference Host
Whitebark Pine Ecosystem Foundation

Organizing Committee

Committee Member	Location	Email
Bob Keane, Program Chair	Missoula, Montana	robert.e.keane@usda.gov
Glenda Scott, Logistics Chair	Missoula, Montana	glenda.scott@whitebarkfound.org
Alina Cansler	Missoula, Montana	alina.cansler@gmail.com
Anna Schoettle	Fort Collins, Colorado	anna.schoettle@usda.gov
Bryan Donner	Columbia Falls, Montana	bryan.donner@whitebarkfound.org
Cara Nelson	Missoula, Montana	cara.nelson@mso.umt.edu
Cathie Jean	Ennis, Montana	terracasa45@gmail.com
Cathy Stewart	Missoula, Montana	cathy.stewart@whitebarkfound.org
Cliff Bara	Troy, Montana	cuda11235@gmail.com
Cyndi Smith	Canmore, Alberta	cyndi.smith@whitebarkfound.org
Danielle Ulrich	Bozeman, Montana	danielle.ulrich@montana.edu
Diana Tomback	Denver, Colorado	diana.tomback@ucdenver.edu
Julee Shamhart	Missoula, Montana	julee.shamhart@whitebarkfound.org
Liz Davy	Ashton, Idaho	liz.davy@whitebarkfound.org
Melissa Jenkins	Bigfork, Montana	melissa.jenkins@whitebarkfound.org
Michael Murray	Nelson, British Columbia	michael.murray@gov.bc.ca
Mike Giesey	Troy, Montana	mike.giesey@whitebarkfound.org
Natalie Stafl	Revelstoke, British Columbia	natalie.stafl@canada.ca
Nicole Hupp	Boulder City, Nevada	nicole_hupp@nps.gov
Shawn McKinney	Missoula, Montana	shawntmckinney@gmail.com
Signe Leirfallam	Wise River, Montana	signe.leirfallom@umontana.edu
Donna Kridelbaugh	virtual	events@whitebarkfound.org

Program Schedule

All times are in Mountain Daylight Time			
Monday, October 4 2021			
10:00 AM - 4:00PM	Crown of the Continent Ecosystem Hi5 Annual Meeting		
Tuesday, October 5 2021			
7:00 AM MDT	Conference Site Available for Log-in		
8:00 - 8:10	**Welcome and Opening Message** Rob Mangold, Director, Whitebark Pine Ecosystem Foundation		
8:10 - 8:40	**Keynote Presentation** Anna Schoettle "Taking the long view: Ecology, condition, and outlook for the high-elevation five-needle pines"		
8:40 - 9:05	**Plenary Presentation** Diana F. Tomback "The National Whitebark Pine Restoration Plan: Restoration model for the high five pines"		
9:05 - 9:30	**Plenary Presentation** Melissa Jenkins "Restoring the Crown: A plan for whitebark pine restoration in the Crown of the Continent Ecosystem"		
9:30 - 10:00	***BREAK*** ***Question/Answer with Keynote and Plenary Speakers*** *- Networking Rooms, select Q&A with Keynote, Plenary 1, or Plenary 2*		
	Room A	Room B	Room C
	1. Historical Whitebark Pine Research Moderator Bob Keane	**2. High-elevation White Pine Communities in the Pacific West Region** Moderator Jonathan Nesmith	**3. The North American Dendroecological Fieldweek (NADEF): Educational and Research Opportunities to Better Understanding 5-needle Pines** Moderator Maegan Rochner
10:00 - 10:15	1-1. My introduction to whitebark pine (Steve Arno)	2-1. Mapping and monitoring California's high-elevation forests: A statewide field and remote sensing campaign (Michele Slaton)	3-1. Education and research through the North American Dendroecological Fieldweek (James H. Speer)

10:15-10:30	1-2. Discovery of WBP, MSU 1974-2001: WBP environment, stand establishment and development, physiology, and follow-up (Tad Weaver)	2-2. An assessment of whitebark pine populations in national parks of the Pacific states (Erik Jules)	3-2. Anatomy measurements from whitebark pine (Pinus albicaulis) in the Beartooth Mountains, WY show strong seasonality signal (April Kaiser)
10:30-10:45	1-3. Seminal research: Foraging strategies of Clark's Nutcracker (Diana Tomback)	2-3. Ecological integrity of whitebark pine on national forests in California (Marc Meyer)	3-3. Summer air temperature for the Greater Yellowstone Ecoregion over 1,250 years (Karen Heeter)
10:45 - 11:00	1-4. The grizzly-squirrel-pine nut connection (Kate Kendall)	2-4. Assessing vulnerabilities in the mutualism between whitebark pine and Clark's nutcracker in Sierra-Cascade parks (Chris Ray)	3-4. Mechanisms of vegetation change in high-elevation forests of the Greater Yellowstone Ecosystem (Erika Blomdahl)
11:00 - 11:15	1-5. Moderated Discussion Panel and Question period with session speakers	2-5. Whitebark Pine Conservation Program at Crater Lake National Park: 2003-2021 (Jen Hooke)	3-5. Variable demographic patterns interact with disturbance to shape limber pine population viability (Justin DeRose)
11:15 - 11:30		2-6. Moderated Discussion Panel and Question period with session speakers	3-6. Whitebark pine as natural archive: the dendrochronological value of whitebark pine and a call to action (Maegen Rochner)
11:30 - 11:45			3-7. Moderated Discussion Panel and Question period with session speakers
11:45 - 12:00			
12:00 - 1:00 PM	*Break*		
	Room A	Room B	Room C
	4. Fire in High-elevation Five-needle Pine Ecosystems Moderator Alina Cansler	**5. High Five Assessments and Analysis** Moderator Shawn McKinney	**6. Progress in Restoration and Management of Five-needle Pines** Moderator Glenda Scott
1:00 - 1:15	4-1. Managing wildfire for whitebark pine ecosystem restoration in Western North America (Bob Keane)	5-1. Limber pine condition on Montana's Rocky Mountain Front (Christy Cleaver)	6-1. Efficacy of pruning limber pine to mitigate white pine blister rust impacts (Kelly Burns)
1:15-1:30	4-2. Whitebark pine regeneration densities, niche, and dynamics three decades after the 1988 Yellowstone Fires (Elizabeth Pansing)	5-2. Application of high-res WorldView-3 satellite imagery to resolve treeline species composition: limber pine in RMNP (Laurel Sindewald)	6-2. Whitebark pine production at the USFS Coeur d' Alene Forest Nursery (Emily Rhoades)
1:30-1:45	4-3. Do whitebark pine regeneration trends support predicted distributional shifts under climate change? Examples from the GYE (Diana Tomback)	5-3. Whitebark & WorldView-2 - looking into the possibilities of using satellite imagery to distinguish subalpine tree species (Stephanie Jouvet)	6-3. Managing high-elevation pine at Great Basin National Park (Gretchen M. Baker)

1:45-2:00	4-4. Fire-caused mortality of whitebark pine (Sharon Hood)	5-4. Assessment of cumulative whitebark pine mortality in the Greater Yellowstone Ecosystem (William Macfarlane | Brian Howell)	6-4. Whitebark pine tree improvement and operational management on the Helena-Lewis and Clark National Forest (Tanya E. Murphy | Matthew O. Voigt)
2:00-2:15	4-5. Fire, restoration and climate change in southwestern mixed conifer forests: implications for P. strobiformis (Kristen Waring)	5-5. Moderated Discussion Panel and Question period with session speakers	6-5. Overview of whitebark pine restoration efforts in the U.S. Forest Service Regions 1, 2, and 4 (Ellen Jungck)
2:15-2:30	4-6. Moderated Discussion Panel and Question period with session speakers		6-6. Limber pine restoration in the Black Hills National Forest, South Dakota (James T. Blodgett)
2:30-2:45			6-7. Moderated Discussion Panel and Question period with session speakers
2:45-3:00			
3:00 - 3:30	***Break - WILD GRACE YOGA with Michelle Geisey (Troy, MT)***		
	Room A	Room B	Room C
	7. Fire in High-elevation Five-needle Pine Ecosystems II Moderator Alina Cansler	**8. Conservation Status, Planning, and Adaptive Management for the High Five Pines I** Moderator Diana Tomback/ Eric Sprague	**9. High Five Pine Ecology** Moderator Bob Keane
3:30 - 3:45	7-1. Does interannual availability of cones affect Clark's Nutcracker caching in burns and habitat selection (Vernon S. Peters)	8-1. Status of whitebark pine: where we go from here (Ben Solvesky)	9-1. Divergent, age associated fungal communities of Pinus flexilis and Pinus longaeva (Joseph Birch)
3:45 - 4:00	7-2. Post-fire regeneration of endangered limber pine (Pinus flexilis) at the northern extent of its range (Denyse Dawe)	8-2. Whitebark pine restoration in BLM Montana's core areas (Emily Guiberson | Rich Byron)	9-2. Alpine treeline ecotones are potential refugia for a montane pine species threatened by bark beetle outbreaks (Colin Maher)
4:00 - 4:15	7-3. Does high severity fire facilitate species transition in California subalpine forests? (Emily Brodie)	8-3. Comparative species assessments of five-needle pines throughout the western United States (Sara A. Goeking)	9-3. A fungal focus: ectomycorrhizal fungi of whitebark pine in interior British Columbia (Hanno Southam)
4:15 - 4:30	7-4. Soil moisture regime and canopy closure structure subalpine understory development over 30 years following stand-replacing fire (Andrew J. Andrade)	8-4. Effective monitoring and evaluation of whitebark pine restoration treatments: key to adaptive management and future applications (Cara R. Nelson)	9-4. Moderated Discussion Panel and Question period with session speakers
4:30 - 4:45	7-5. Moderated Discussion Panel and Question period with session speakers	8-5. Moderated Discussion Panel and Question period with session speakers	
4:45 - 5:00			
5:00 - 5:30 pm	***Break and transition to poster session***		

5:30 - 6:30 pm	**"Posters, Pines, and Wine"** (***Self-guided Virtual Poster Session - visit with Authors***)
7:00 - 8:00 pm	**High Five Pines and Communities They Bind** (***Virtual Event- Community invited***)

Wednesday, October 6 2021			
7:00 AM MDT	Conference Site Available for Log-in		
8:00 - 8:10	**Welcome**		
	Eric Sprague, Director of Forest Conservation, American Forests		
8:10 - 8:40	**Keynote Presentation**		
	Richard Sniezko "Genetic resistance to white pine blister rust, restoration options, and potential use of biotechnology"		
8:40 - 9:05	**Plenary Presentation**		
	David Neale "The Whitebark Pine Genome Project"		
9:05 - 9:30	**Plenary Presentation**		
	Bob Keane "Effective actions for managing resilient high-elevation five-needle pines in western white pine forest of North America under changing climates at multiple scales"		
9:30 - 10:00	***BREAK*** ***Question/Answer with Keynote and Plenary Speakers*** - *go to Networking Room, select Q&A with Keynote, Plenary 1, or Plenary 2*		
	Room A	Room B	Room C
	10. Ecophysiology of Five-Needle Pines Moderator Danielle Ulrich	**11. Five-Needle Pine Research, Monitoring, and Restoration in National Parks** Moderator Kristin Legg	**12. High Five Restoration and Management I** Moderator Cara Nelson

10:00 - 10:15	10-1. A tale of two pines: comparing seedling morphological and stomatal traits between whitebark pine and limber pine (Meghan McFadden)	11-1. Adaptive monitoring in action: reconsidering design-based estimators reveals underestimation of whitebark pine disease prevalence in the Greater Yellowstone Ecosystem (Erin Shanahan | Kathi Irvine)	12-1. Restoration planting options for limber pines (Pinus flexilis *James*): 10 years later (Anne Marie Aramati Casper)
10:15 - 10:30	10-2. Within and between species physiological traits and stress resistances of three high-elevation pines (Chloe Wasteneys)	11-2. Compounding effects of white pine blister rust, mountain pine beetle, and fire on white pines in Sequoia and Kings Canyon National Parks (Jonathan Nesmith)	12-2. Forest structure twenty years after the first whitebark pine prescribed burn in Banff National Park (Brendan Wilson)
10:30 - 10:45	10-3. Whitebark pine (Pinus albicaulis) growth and defense in response to mountain pine beetle outbreaks (Nickolas Kichas)	11-3. What have we learned in 20 years of whitebark pine restoration in Glacier National Park? (Rebecca Lawrence | Jennifer Hintz Guse)	12-3. Monitoring the effects of prescribed burning on whitebark pine (Cara Nelson)
10:45 - 11:00	10-4. Differential spring budburst phenology across western five-needled pine species (Franklin Alongi)	11-4. Great Basin bristlecone and limber pine in Great Basin National Park: first look into the long- term white pine monitoring efforts (Nicole Hupp | Jeff Galvin)	12-4. Long-term assessment of the efficacy of prescribed burning and mechanical thinning for restoration of whitebark pine (Enzo Martelli)

11:00 - 11:15	10-5. Physiological responses of whitebark, limber and Great Basin bristlecone pines to environmental stress (Sean Hoy-Skubik)	11-5. Climatic factors affecting white pine blister rust incidence on whitebark pine in Washington National Parks (Sebastian Espinosa Novoa)	12-5. Verbenone and green-leaf volatiles reduce whitebark pine mortality in a northern range-expanding mountain pine beetle outbreak (Brenda Shepherd)
11:15 - 11:30	10-6. Winter damage is more important than summer temperature for maintaining the krummholz growth form above alpine treeline (Colin Maher)	Challenges to limber pine at Craters of the Moon National Monument and Preserve (Devin Stucki)	12-6. Moderated Discussion Panel and Question period with session speakers
11:30 - 11:45	10-7. Moderated Discussion Panel and Question period with session speakers	11-6. Moderated Discussion Panel and Question period with session speakers	
11:45 - 12:00			
12:00 - 1:00 PM	*Break*		
12:15 to 12:45	**Whitebark Pine Ecosystem Foundation Members Meeting- Open Invitation**		
1:00 - 3:10	Virtual Field Trip to the Smith Creek Research Site		
	Welcome and Logistics - Bryan Donner		
	The Smith Creek Research Area: History and Objectives - Cathy Stewart, Leslie Anderson, and Bob Keane		
	Restoration strategies – local, regional, and range-wide strategies for restoration - Diana Tomback		

	Interacting disturbances: how insect, fire, and disease act to influence high-elevation pine dynamics under climate change - Barbara Bentz	
	Smith Creek Lessons Learned - What actually happened that was not anticipated twenty-two years ago - Bob Keane and Steve Arno	
	Wildland fire: High-elevation fire regimes, wildland fire management, and climate change - Alina Cansler	
	Management challenges: regulations, laws, and oversight of local to international management of high mountain resources - Randy Moody	
	Restoration Strategies on the Bitterroot National Forest, Montana - Cheri Hartless	
	Discussion Panel -Q&A with all the presenters	
3:10-3:30	*Break*	
3:30 - 5:00	**Student and Early Professionals Roundtable**	**Networking** (concurrent activities)
	Moderator Cara Nelson	**Networking rooms open with focused discussions**
3:30 - 3:35	Introductions	1- Rust Ecology and Resistance
3:35-4:15	**Panel: "Federal Hiring: the Ins and Outs of Getting Hired by the US Forest Service and National Park Service"**	2- Restoration and Management
	Amber Kamps, USFS Regional Outreach, Recruitment, & Retention Program Specialist, FS Region 1	3- Wildland Fire and Climate Change
	Laura Chavers, NPS Supervisory Human Resources Specialist, Interior Regions 6, 7 and 8	4- NWPRP: National Whitebark Pine
	Questions and Discussion period	5- Wilderness Issues in High Five Pine
4:15 - 4:55	**Panel: "Getting your Data Published"**	**Exhibit Hall open** Posters, Sponsor and Exhibitor Booths
	Bob Keane, Editor of International Journal of Fire Ecology	**Individual interaction opportunities**
	Stephen Murphy, Editor-in-Chief of Restoration Ecology Journal	
	Question and Discussion	
4:55 - 5:00	Closing discussion	
5:00-6:30	*Break*	

6:30 - 8:00	High Five Social (*Virtual Event for Conference Attendees*)

Thursday, October 7 2021

7:00 AM MDT	Conference Site Available for Log-in
8:00 - 8:10	**Welcome** Diana Tomback, Board of Directors, Whitebark Pine Ecosystem Foundation
8:10 - 8:40	**Keynote Presentation** Charlie Cartwright "Early results from whitebark pine genecology trials in British Columbia"
8:40 - 9:05	**Plenary Presentation** Andrew Hansen "Potential changes in climate suitable habitat of Whitebark pine in Greater Yellowstone under climate scenarios"
9:05 - 9:30	**Plenary Presentation** Barbara Bentz "Defense characteristics among high-elevation Pinus and vulnerability to native bark beetles"
9:30 - 10:00	***BREAK*** ***Question/Answer with Keynote and Plenary Speakers-*** *go to Networking/ Networking Room, select Q&A with Keynote, Plenary 1 or Plenary 2*

	Room A	Room B	Room C
	16. High Five Pine Ecology and Management II Moderator Liz Davy	**17. White Pine Blister Rust Resistance, Genetic Variation, and Restoration of White Pines I** Moderator Rich Sniezko	**18. Conservation Status, Planning, and Adaptive Management for the High Five Pines II** Moderators Diana Tomback & Eric Sprague
1:00 - 1:15	16-1. High-elevation five-needle pine seedling traits vary according to climatic gradients (Lacey Hankin)	17-1. Field-testing whitebark pine resistance to white pine blister rust: a cost-effective approach to progeny testing (Iain R. Reid)	18-1. Integrating restoration practices for whitebark pine with climate change distributional shifts (Katie Ireland)
1:15 - 1:30	16-2. Can we improve western white pine microbiomes to promote resistance to blister rust disease? (Lorinda S. Bullington)	17-2. Evaluating spectroscopy as a tool for rapid identification of WPBR-resistant whitebark pine trees (Pierluigi Bonello)	18-2. Conservation and restoration of whitebark pine by the US Forest Service in Oregon and Washington (Andrew Bower)
1:30 - 1:45	16-3. Does drought response change with parental ancestry across a white pine hybrid zone? (Lulu Peach)	17-3. White pine blister rust resistance programs for Alberta limber and whitebark pine recovery (Jodie Krakowski)	18-3. Whitebark pine restoration on the Flathead Indian Reservation: Tribal cultural benefits reaped from new management strategies (ShiNaasha Pete)

1:45 - 2:00	16-4. Ecology and distribution of Klamath foxtail pine (Michael Edward Kauffmann)	17-4. Moderated Discussion Panel and Question period with session speakers	18-4. High-elevation chilgoza pine forests in western Himalayas: survival threats and conservation exigencies (Rinki Sarkar)
2:00 - 2:15	16-5. Moderated Discussion Panel and Question period with session speakers		18-5. Moderated Discussion Panel and Question period with session speakers
2:15 - 2:30			
2:30 - 3:00	***Networking Break***		
	Room A	Room B	
	19. The Myths and Realities of Whitebark Pine and Climate Change Moderator Bob Keane	**20. White Pine Blister Rust Resistance, Genetic Variation, and Restoration of White Pines II** Moderator Rich Sniezko	
3:00 - 3:15	19-1. Predicted impacts of climate change on the northern range limit of whitebark pine (Alana J. Clason)	20-1. Whitebark pine restoration at Crater Lake National Park: the first ten years (Jen Hooke)	
3:15- 3:30	19-2. Nonlinear shifts in white pine blister rust due to climate change (Joan C. Dudney)	20-2. Range-wide testing of white pine blister rust resistance in foxtail pine (Pinus balfouriana) (Richard A. Sniezko)	

Poster Hall

Poster Title	Presenter
Teaching students about whitebark pine and fire using hands-on activities	Ilana Abrahamson
Challenges of restoring a unique population of whitebark pine	Robin Garwood | Deb Taylor
Whitebark pine friendly ski areas	Mike Giesey
Genetic and remote sensing approaches to identify white pine blister rust infection in southwestern white pine.	Jeremy Johnson
Identifying patterns of blister rust resistance in southwestern white pine	Jeremy Johnson
Repeat photography tells whitebark pine stories	Jane Kapler Smith
Pinus flexilis is highly a highly susceptible host of *Dothistroma septosporum* in Canada- first report	Jodie Krakowski
Preparing for invasion: rust resistance in limber, Great Basin bristlecone, and Rocky Mountain bristlecone pines	Anna W. Schoettle
The great basin five-needle pine proactive strategy engagement: a collaborative and science-based approach	Anna W. Schoettle
Climate correlates of white pine blister rust infection in whitebark pine in the greater Yellowstone ecosystem	David Thoma/Erin Shanahan
Conifer regeneration hinders digital estimation of understory plant cover in post-fire whitebark pine communities	Brandi E. Wheeler
Maintaining *Pinus strobiformis*, a tree species threatened by climate change and white pine blister rust	Nicholas Wilhelmi

Variable Demographic Patterns Interact With Disturbance and Climate to Shape Limber Pine Population Viability

R. Justin DeRose, <justin.derose@usu.edu>, Department of Wildland Resources and Ecology Center, Utah State University, Logan UT; **Zachary Merrill**, College of Forest Resources and Environmental Science, Michigan Technological University, Houghton, MI; **Maegen Rochner**, Department of Geographic and Environmental Sciences, University of Louisville, Louisville, KY; **Matthew F. Bekker**, Department of Geography, Brigham Young University, Provo, UT.

ABSTRACT

Limber pine (*Pinus flexilis*) is an important five-needle pine that has received far less research attention than its close taxonomic relative, whitebark pine (*Pinus albicaulis*). Recent, severe, and widespread decline of whitebark pine portends a similar fate for limber pine. While mountain pine beetle (*Dendroctonus ponderosae*) outbreaks are likely responsible for the vast majority of limber pine mortality, their possible role as disturbance agents on demographic processes in these forests is unknown. In this study, we examine the long-term persistence of a limber pine woodland located in the Greater Yellowstone Ecosystem. Dendroecological methods were employed to reconstruct establishment dates, death dates, cause of death, and the driver of growth variability for 81 limber pines. Over the past ~800 years, recruitment was characterized as episodic, with peaks occurring in the 13th and 17th centuries consistent with multi-decadal pluvial periods. Over the past two centuries, mortality was much more continuous, but with pulses centered on the 1880s, 1960s, and early 2000s. Mortality of over half the limber pine since ~1860 can be directly attributed to the mountain pine beetle. Mortality peaked during the 1960s, but was found in trees up to 800+ years old. Variability in limber pine growth was strongly and consistently driven by spring moisture availability. The balance between continuous and episodic demographic processes in the face of mounting disturbance pressure over time is likely to further bottleneck the population, potentially decreasing limber pine population viability.

Keywords: climate change adaptation, generalist, Greater Yellowstone Ecosystem, limber pine, mountain pine beetle, *Pinus flexilis*, population dynamics, surrogate species

INTRODUCTION

Limber pine (*Pinus flexilis*) exhibits such a wide ecological amplitude that it could be considered a poster-child for generalist species. It has a huge geographic range, and unlike any of the other five-needle pines it occurs across the elevation gradient of forests in the interior western United States in proportion with its common associates (Windmuller-Campione and Long 2016). Limber pine's close taxonomic relative, whitebark pine (*Pinus albicaulis*), is commonly cited as a species in peril (USFWS 2020), largely due to combined impacts from the native mountain pine beetle (*Dendroctonus ponderosae*), the invasive white pine blister rust (*Cronartium ribicola*), and compounding influences of climate change (Resler and Tomback 2008; Rochner et al. 2021). While broadly more generalist than whitebark pine, limber pine is not immune to these threats. Indeed, the species is highly vulnerable to both the mountain pine beetle and white pine blister rust. Recent work has indicated that range-wide mortality in limber pine is approaching similar levels as was seen in whitebark pine two decades ago due to mountain pine beetle_(Goeking and Windmuller-Campione 2021). Although limber pine plays a similar functional role as whitebark pine in high-elevation ecosystems, it has received much less research interest (Means 2011; Schoettle et al. 2019). To address this, we examine patterns of establishment, growth, and mortality in a limber pine woodland.

The primary dispersal mechanism for limber pine is seed collectors (Hutchins and Lanner 1982), specifically the Clark's nutcracker (*Nucifraga columbiana*), which likely play a large role in the massive ecological amplitude (Lanner and Vander Wall 1980). However, the establishment and survival of cached seeds depend on the autecological characteristics of limber pine. While limber pine is intolerant of shade, its tolerance to other environmental stressors is high (Steele 1990), such that seed caches in the open areas, such as recently disturbed sites, are more likely to establish and recruit to the overstory. This may be why limber pine is often cited as a post-fire invader; however, fire does not guarantee establishment (Dawe et al. 2020). Limber pine can also be incredibly long-lived: It is not uncommon to find lower elevation trees in excess of 700 years, and higher elevation trees in excess of 1,000 years old (MacDonald and Case 2005). Indeed, some remarkable limber pine (live and dead) exist in the west, preserved by a combination of their longevity (high resin content, exposed sites), and also their remoteness.

To better understand the potential for limber pine population decline or persistence we sampled a putative limber pine "ghost woodland" within the Greater Yellowstone Ecosystem that had what appeared to be substantial levels of mortality (figure 1). We used dendroecological techniques to reconstruct establishment dates, mortality dates, and drivers of growth variability. While this limber pine population seems to have been threatened by the recent mountain pine beetle outbreak that began in the early 2000s, not unlike that seen in whitebark pine, perhaps this kind of disturbance is a

Figure 1. Pictures of the Bald Ridge study area: (a) view of the limber pine woodland and ridge extent, (b) example of live limber pine, (c) western edge of the study area, (d) & (e) sampling dead limber pine.

normal part of population dynamics. To address this possibility, we ask the following questions: 1) Were dates of mortality continuous or episodic, and was the recent mountain pine beetle outbreak the primary cause? 2) Was reconstructed recruitment continuous or episodic? 3) What climatic factors influenced growth variability, were they temporally stable, and were establishment or mortality patterns driven by these climatic factors?

METHODS

Study Area and Sample Selection

Bald Ridge is a gently east-sloping fault block with a relatively open canopy woodland of limber pine (44.789°N, 109.338°W, 2330 m elevation), in the Shoshone National Forest near Cody, WY (figure 1). Substantial mortality was apparent at this site, putatively caused by the recent mountain pine beetle outbreak. This site is south-facing with minimal soil development. To capture the influence of climate on tree growth, we sought trees from this site that were greater than 10 cm in diameter at breast height and appeared to be, or have been, open grown, such that we would minimize the effects of possible density-dependent competition on growth.

Tree-Ring Data Collection and Chronology Development

We collected two cores per tree, roughly perpendicular to each other at or near ground level but sometimes resorted to a higher point on the stem (up to 60 cm) because of root buttressing, branches, or other obstructions. We also recorded tree status (live or dead), coring height, diameter at coring height, and GPS location. In addition to cores, we also collected full cross-sections near the root collar of snags and sub-fossil wood. We mounted, sanded, and dotted the samples using standard procedures (Speer 2010). While dotting the samples, we used visual crossdating methods, which allowed for the identification of potential dating errors and missing rings prior to measurement and statistical crossdating. We measured tree-ring widths with a Velmex Unislide Measure J2X system with 0.001 mm precision. To verify our crossdating we used COFECHA (Holmes 1983). We verified the annual resolution of every tree reported in our results through crossdating. To develop a site chronology for climate response we detrended each series using a cubic smoothing spline, averaging with a robust mean, and applying an autoregressive model in *dplR* (Bunn 2008). We used the residual chronology in all climate-growth analyses.

Establishment Dates and Death Dates

We were confident we could estimate establishment dates for many of the limber pine samples despite a tendency for heart rot in the species. To estimate the number of rings to the pith for all samples that did not reach tree center, we used either the Duncan (1989) or the Applequist (1958) methods for cores and cross-sections, respectively. We chose the best estimates (i.e., oldest and most confident pith date) made from all measured series for each sample (two per tree in most cases) and binned these into bi-decadal groups for graphical analysis.

We recorded dates of death for any limber pine in which we had strong evidence for the waney edge, that is, that the outside ring was the last year of growth, indicated by either smooth outer wood or the presence of bark (figure 2). We noted how confident we were in the outside ring being indicative of the date of death. We chose the best estimate (i.e., latest, and most confident) made from all measured series for a given tree (two per tree in most cases). We also noted the presence or absence of blue-stain fungi in the entirety of the present sapwood of the dead samples, when sapwood was present, as a proxy for indicating that the tree was putatively killed by the mountain pine beetle.

Climate Growth Relationships

We evaluated climate-growth relationships between the residual chronology and monthly precipitation and maximum monthly temperature. Climate data extracted from the Partial Regression on Independent Slopes Model dataset (PRISM 2018) was used in the R package *treeclim* (Zang and Biondi 2015). We evaluated monthly and seasonal (3, 6, and 12-month) summed (precipitation) or averaged (temperature) responses under the assumption that precipitation was the primary driver, and temperature the secondary, based on substantial preliminary testing (Meko et al. 2011). We then tested for temporal stability in response to the primary growth driver using an evolving correlation analysis that started with a 30-yr moving window that increased by one year in each iteration with significance at the $P < 0.05$ level based on bootstrapped confidence bounds (Biondi 1997).

RESULTS

We crossdated and measured 120 tree-ring series from 63 trees sampled on Bald Ridge to build an initial limber pine chronology. The inter-series correlation was 0.561, and

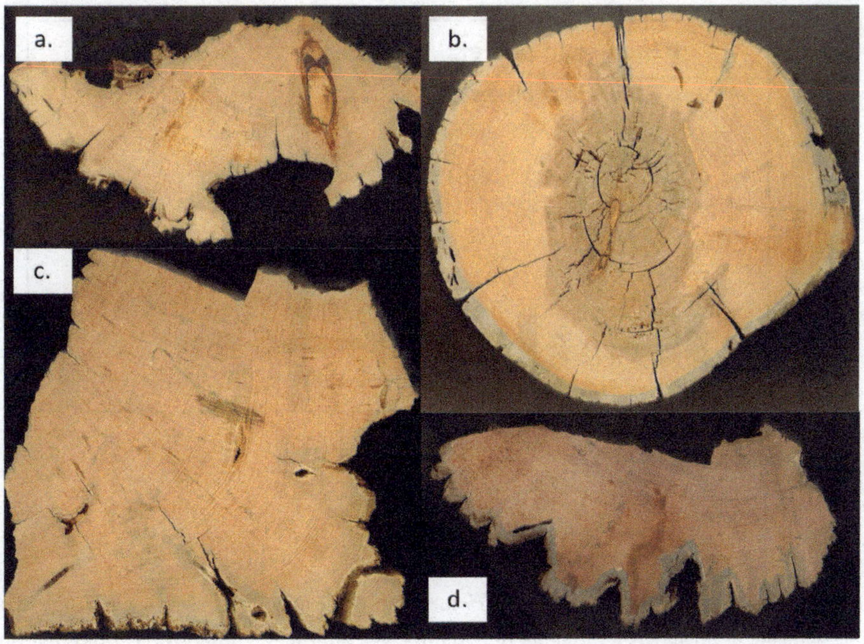

Figure 2. Examples of limber pine cross-sections. (a) Cross-section without indication of outside ring or sapwood, but inside curvature sufficient for the determination of an establishment date. (b) More recently killed tree with a waney edge, blue-stain in the sapwood, and a pith. (c) Very old cross-section with no outside ring, no sapwood, and no way to determine establishment date. (d) Very old cross-section with no outside edge, weathering that could be mistaken for sapwood, and no way to determine establishment date.

average standard deviation was 0.285. Mean length for all series was 307 years with a total of 36,901 rings measured. We used this chronology to crossdate 18 additional trees, for 81 trees in total. Of those, 30 were alive, and 51 were dead. Fifty-three percent of the trees contained the pith (n = 43, five live and 38 dead trees). Twenty-four trees without the pith had curvature sufficient to estimate the number of rings to the pith. Of the dead trees with evidence of the waney edge, we confirmed date of death for 18 limber pines. The

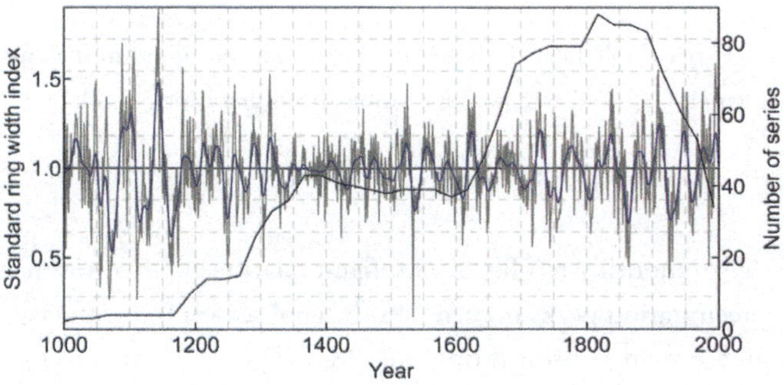

Figure 3. Standard limber pine chronology for Bald Ridge plotted to accentuate year-to-year and multi-decadal variability in ring width. Dark blue line is a 20-year cubic smoothing spline. Secondary axis indicates the number of tree-ring series in the chronology.

full limber pine chronology for Bald Ridge covered the period (965-2018, figure 3).

We obtained confirmed mortality dates for only 2% (n = 18) of the total limber pine sampled. A large majority of the dead limber pine with confirmed death dates (77%; 14 of 18 trees) died from mountain pine beetle attack, presumed by the presence of blue-stain fungi in the sapwood. While the mountain pine beetle-caused mortality peaked in the 1960s, its presence in the sapwood of limber pine across the site at Bald Ridge suggested it has been a disturbance agent in this system continuously for the last millennium (figures 4 and 5).

We estimated establishment dates from 81% (n = 66) of the limber pine sampled. The average estimated number of rings to the pith was 22.3. The oldest establishment date was 965, and the most recent occurred in the 1940s (figure 5). The oldest live tree establishment date was 1648. Over the past approximate millennium, establishment of limber pine at Bald Ridge was characterized by pulsed recruitment primarily in the 1300s, and from the 1600s through the mid-1700s (figure 5). Of the trees used to estimate establishment dates, 28 (42%) had evidence of blue-stain in the sapwood. None of the live trees had blue-stain in their sapwood.

Variability in limber pine ring width at Bald Ridge was positively correlated with precipitation over the historical in-

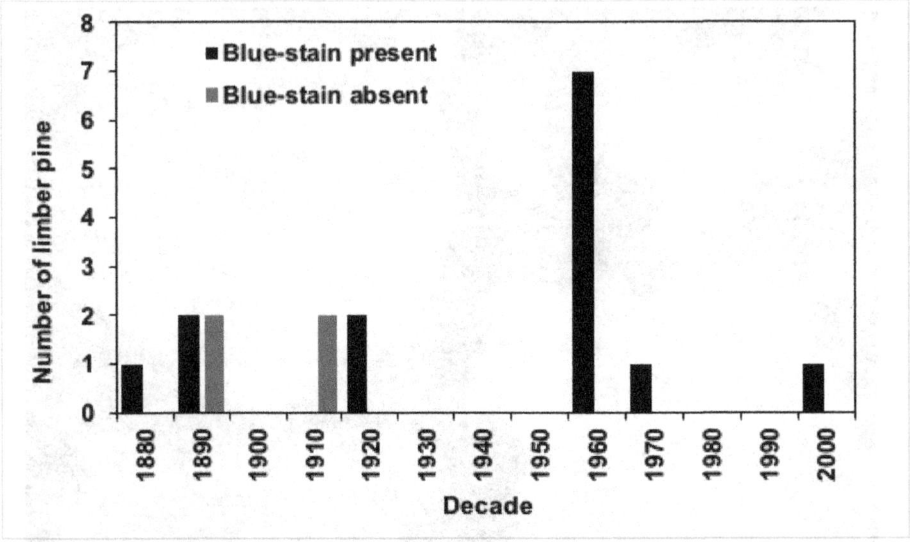

Figure 4. Number of limber pine by decadal bin that have a confirmed date of death. Black-filled bars indicate that blue-stain was found in the sapwood of the sample, while gray-filled bars indicate no blue-stain. Gaps indicate no observations from that time period.

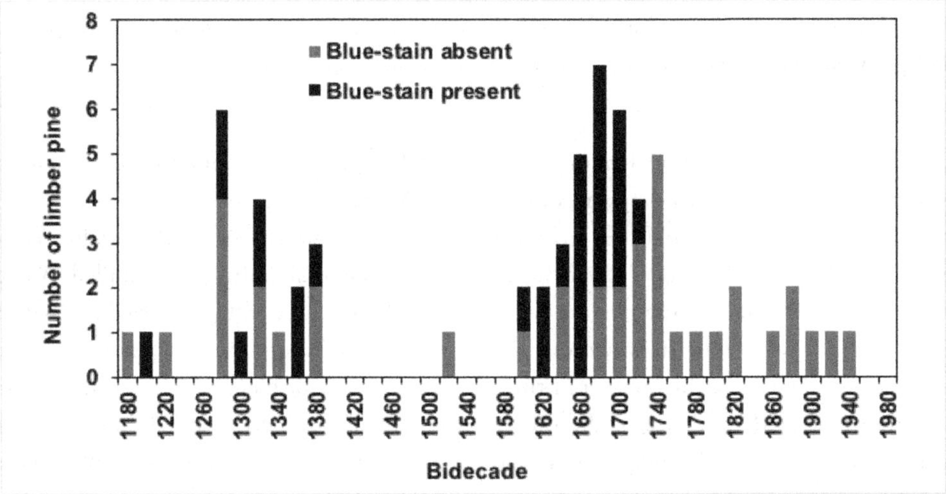

Figure 5. Number of limber pine established by bidecadal bin at Bald Ridge. Black-filled bars indicate that blue-stain was found in the sapwood of the sample, while gray-filled bars indicate no blue-stain or no sapwood. Gaps indicate no observations from that time period.

strumental record (1896-2017). After variability in growth due to precipitation was accounted for, temperature explained no further variation (figure 6). Month-by-month limber pine growth could be explained by growing season precipitation during May, and cool season January precipitation (i.e., snowpack). Seasonal accumulation of precipitation explained more variability in limber pine growth, specifically the water-year ending in May of the growing season (figure 6). Similar to monthly results, the strongest relationships to seasonal precipitation peaks occurred in the cool season (i.e., January) and growing season spring (i.e., May; figure 7). The positive relationship between limber pine growth and monthly May precipitation was statistically significant ($P < 0.05$) and temporally stable over the historical record of climate (figure 7). However, growth relationships with cool-season precipitation were not temporally stable over the same period; we identi-

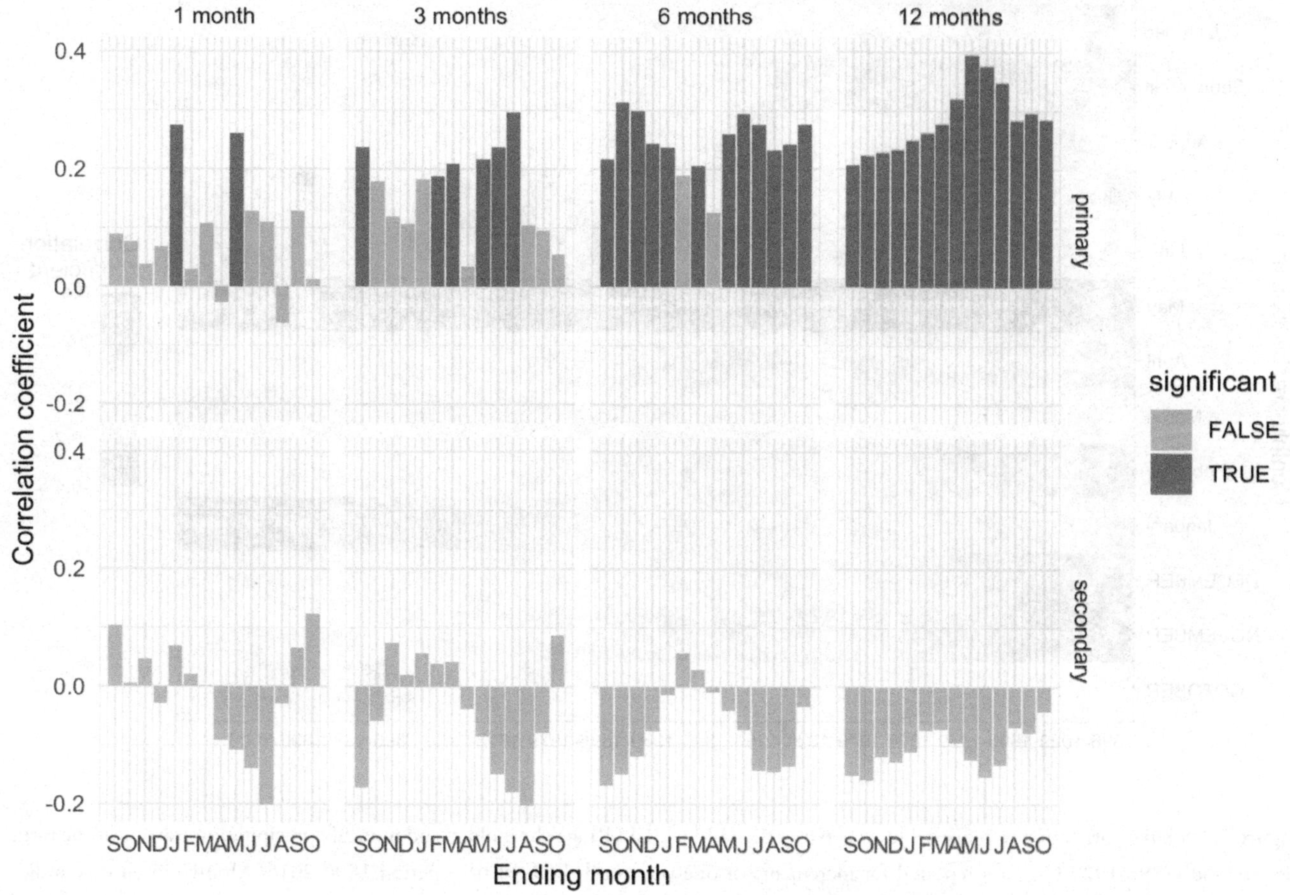

Figure 6. Monthly and seasonal (3, 6, and 12-month) correlations between the Bald Ridge residual chronology and precipitation as the primary variable (top panels), and maximum temperature data as the secondary variable (bottom panels). Significance at the 95% confidence level is indicated by shaded bars.

fied a shift from a negative cool-season response (previous November and December) to a positive cool-season response (current January) during the mid-20th century (figure 7).

DISCUSSION

For at least the last millennium a limber pine woodland has occupied the Bald Ridge site. Over this period episodic establishment, coupled with continual mortality and moisture-driven growth, has resulted in relatively continuous population dynamics. Although our data were not sufficient to ascertain whether the mountain pine beetle-caused mortality had increased in recent decades, it is clear mountain pine beetle has been killing limber pine on Bald Ridge for as long as we have been able to infer based on the presence of blue-stain fungi. One possibility for the lack of recent mountain pine beetle-killed limber pine was our sampling approach which could have been inadvertently focused on older specimens (i.e., trees larger than 10 cm in diameter at breast height), and therefore created a bias against recently dead trees. Regardless, background mortality rates were very low, averaging 1.6% per decade over the relatively short period of reconstructed death dates (~1880-2000). The visible presence of standing dead and down limber pine in this "ghost woodland," begs the question, was the observed mortality a normal part of disturbance processes in limber pine systems, or does it portend a more challenging future for the species?

By far the biggest contributor to the observed mortality at Bald Ridge has been the mountain pine beetle, but this disturbance agent has been around for millennia (Brunelle et al. 2008) at endemic and epidemic levels. However, outbreaking mountain pine beetle populations might be increasingly

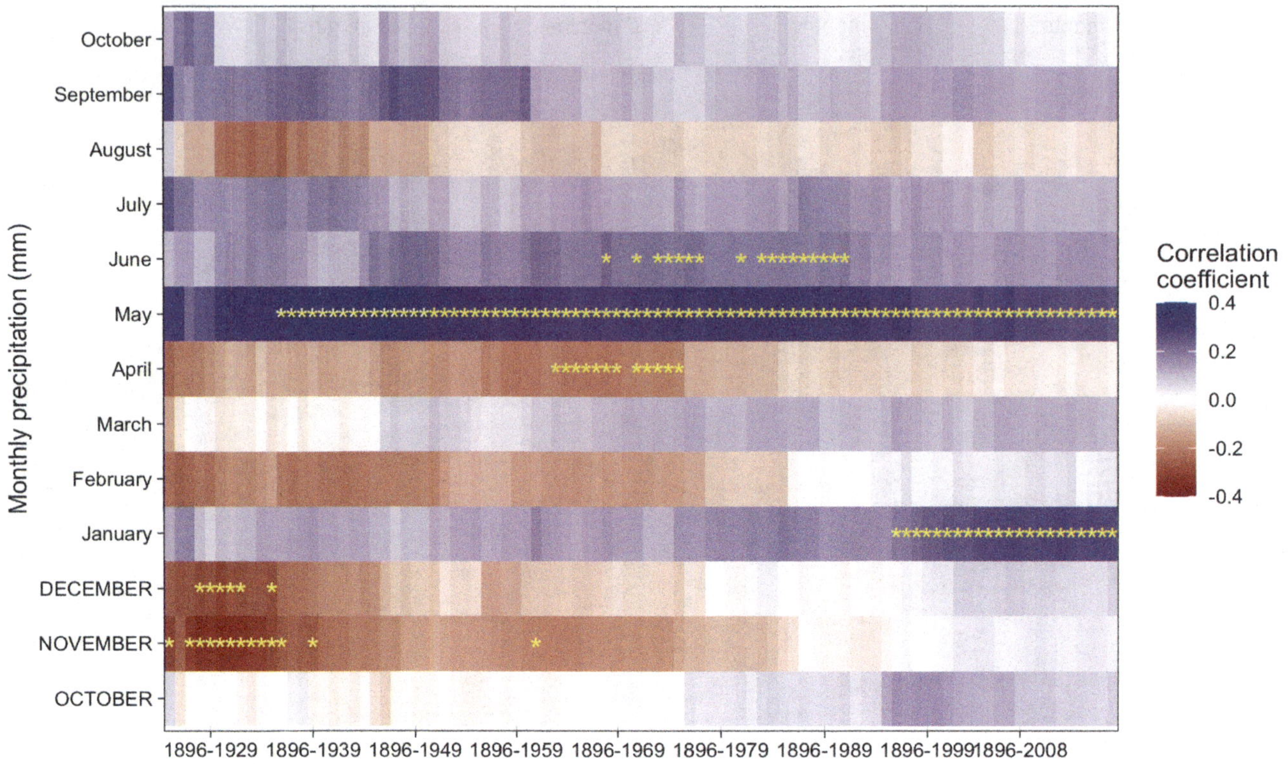

Figure 7. Evolving correlation response window between the residual Bald Ridge chronology and monthly precipitation starting at the earliest time period (1896-1929 CE) and repeated for increments of one year until the final time period 1896-2015. Months in all caps indicate previous year. Yellow stars indicate significance of correlation for the evolving response window at the p<0.05 level.

likely at higher elevations due to climate change (Raffa et al. 2008), but limber pine exists across all elevations in its geographic range where climate may not have been historically limiting for the mountain pine beetle. If this is the case, it may help explain why limber pine at Bald Ridge have been affected by mountain pine beetle for centuries. Despite the susceptibility of limber pine to mountain pine beetle, older trees, and survivors of mountain pine beetle outbreaks, do persist on harsh sites.

Because climate-growth analyses indicated that ring width variability at Bald Ridge responded to seasonal precipitation, it can serve as a moisture proxy going back in time. In contrast to continuous mortality, largely from mountain pine beetle, pulses of successful establishment appear to have been coincident with wet climatic periods (figures 3 and 7), at least during some decades, despite assumed similar levels of pressure by Clark's nutcracker and other seed predators over time. While there may have been limber pine establishment since the 1940s, our sampling approach (increment cores) would have precluded trees too young (small) to extract cores, and might partially explain limited establishment in the previous ~80 years. Similarly, although we were able to reconstruct pith dates for over 80% of the trees sampled, patterns in establishment dates may differ had we captured a higher percentage of the population.

Moisture conditions were the primary driver of tree growth; water-year (12-month), particularly spring precipitation, was the dominant climatic factor influencing growth variability at the Bald Ridge site. We found that drought stress (high temperature and low precipitation) limits growth in limber pine, which was consistent with previous studies (Kipfmueller and Salzer 2010). The changing cool season precipitation signal suggests more precipitation is being delivered as rain, rather than snow over time, providing less of a potential negative effect of late-lying snowpack on growth, and increased opportunity for soil moisture to contribute to enhanced growth. One of the primary factors in tree growth, photosynthetic capacity, is responsive to moisture availability, and higher temperatures limit respiration rates in trees. It is likely that the generally exposed nature of our Bald Ridge site resulted in increased solar radiation over the course of the growing season that would effectively control the availability of soil moisture, thereby limiting annual ring-width increment. As we look to the future of limber pine, these observations will be important to consider as we head into a time period with likely more drought, increased temperatures, and less frequent and/or more variable wet periods (Westerling et al. 2011).

Overall, the balance between continuous mortality, episodic establishment, and climate driven growth variability in the face of mounting mountain pine beetle pressure and increasing precipitation variability is likely to impinge upon limber pine population viability. Introduced pathogens, specifically white pine blister rust, also threaten limber pine forest health, although there is some indication of genetic resistance (Sniezko et al. 2016). Still, the existence of many persistent and old limber pines at Bald Ridge indicate the species is likely to persist, yet. These longer-lived, slower-growing specimens may also represent the seed source for future generations of limber pine as climate changes. Across a broad ecological amplitude, the presence of many live trees, including such long-lived survivors, could make limber pine populations a resource for maintaining the function of five-needle pines in the face of rapid environmental change. Our dendroecological results constitute a better understanding of limber pine ecology, including the mechanisms behind survival and mortality of individuals, that would benefit management of the Greater Yellowstone Ecosystem and other montane and subalpine limber pine forests in the interior west.

Using dendrochronological methods with limber pine is not without its limitations. It is widely known that individuals of advanced age (often the most sought-after specimens) are likely to be rotten in the center, making estimates of establishment dates difficult. Our intensive sampling approach was able to remedy this as we ended up with quite a large number of samples with the pith or near pith growth rings, enabling us to make a reasonable estimate of the number of rings to the pith (figure 2). Given the potential for long tree-ring chronologies of limber pine, it is surprising there are so few efforts to build reconstructions with this species (but see Case and MacDonald 1995; Axelson et al. 2009). While often characterized as variable in their climate signal (Kipfmueller and Salzer 2010), limber pine can be excellent and stable recorders of past climatic variability. Their ecological amplitude, longevity, and landscape placement all make them ideal candidates for the development of long tree-ring chronologies. Such chronologies, as seen with our study, can provide useful information on long-term demographics, disturbance, and climate drivers in limber pine populations.

This particular study aids in improving our understanding of limber pine ecology, and can be used to guide our management of the species into the future. As we think about climate change adaptation, and the conservation, preservation, or restoration of five-needle pines, we suggest that, as a textbook generalist, limber pine has untapped potential to help us maintain five-needle pines on our western landscapes. Whether that is the promotion of extant populations, the release of pine from more successful overstory competitors, or its use as a 'surrogate' to fill recently vacated niches of functionally similar species (e.g., whitebark pine), we need to look more closely at the adaptation potential of limber pine.

ACKNOWLEDGEMENTS

This paper was a product of the 2018 North American Dendroecological Fieldweek, conducted at the A.L. Mickelson Field Station, Cody, Wyoming, and we wish to thank the organizers. We want to thank Charles Gertler, Julia Schwager, and Samuel Wyatt for help in the field and lab and early comments on this paper. We especially want to thank Ryan Jess for in-depth identification of sapwood, blue-stain, and pith offsets in the lab. This research was supported by the Utah Agricultural Experiment Station, Utah State University, and approved as journal paper number 9535. We wish to thank Jodi Axelson and Kristen Pelz for their constructive comments which greatly improved this manuscript.

LITERATURE CITED

Applequist, MB. 1958. A simple pith locator for use with off-center increment cores. Journal of Forestry 56:141. doi.org/10.1093/jof/56.2.138.

Axelson, JN, DJ Sauchyn, and J Barichivich. 2009. New reconstructions of streamflow variability in the South Saskatchewan River Basin from a network of tree ring chronologies, Alberta, Canada. Water Resources Research 45:W09422. doi.org/10.1029/2008WR007639.

Biondi, F. 1997. Evolutionary and moving response functions in dendroclimatology. Dendrochronologia 15:139-150.

Brunelle, A, GE Rehfeldt, B Bentz, and AS Munson. 2008. Holocene records of *Dendroctonus* bark beetles in high-elevation pine forests of Idaho and Montana, USA. Forest

Ecology and Management 255:836-846. doi.org/10.1016/j.foreco.2007.10.008.

Bunn, AG. 2008. A dendrochronology program library in R (dplR). Dendrochronologia 26:115-124. doi.org/10.1016/j.dendro.2008.01.002.

Case, RA, and GM MacDonald. 1995. A dendroclimatic reconstruction of annual precipitation on the western Canadian Prairies since A.D. 1505 from *Pinus flexilis* James. Quaternary Research 44:267-275. doi.org/10.1006/qres.1995.1071.

Dawe, DA, VS Peters, and MD Flannigan. 2020. Post-fire regeneration of endangered limber pine (Pinus flexilis) at the northern extent of its range. Forest Ecology and Management 457(5): 117725. doi.org/10.1016/j.foreco.2019.117725.

Duncan, RP. 1989. An evaluation of errors in tree age estimates based on increment cores in kahiketea (*Dacrycarpus dacrydioides*). New Zealand Natural Sciences 16:31-37.

Goeking, SA, and MA Windmuller-Campione. 2021. Comparative species assessments of five-needle pines throughout the western United States. Forest Ecology and Management 496:119438. doi.org/10.1016/j.foreco.2021.119438.

Holmes, RL. 1983. Computer-assisted quality control in tree-ring dating and measurement. Tree-Ring Bulletin 43:69-78.

Hutchins, HE, and RM Lanner. 1982. The central role of Clark's nutcracker in the dispersal and establishment of whitebark pine. Oecologia 55:192-201. doi.org/10.1007/BF00384487.

Kipfmueller, KF, and MW Salzer. 2010. Linear trend and climate response of five-needle pines in the western United States related to treeline proximity. Canadian Journal of Forest Research 40:134-142. doi.org/10.1139/X09-187.

Lanner, RM, and SB Vander Wall. 1980. Dispersal of limber pine seed by Clark's nutcracker. Journal of Forestry 78:637-639. doi.org/10.1093/jof/78.10.637.

MacDonald, GM, and RA Case. 2005. Variations in the Pacific Decadal Oscillation over the past millennium. Geophysical Research Letters 32:L08703. doi.org/10.1029/2005GL022478.

Means, RE. 2011. Synthesis of lower treeline limber pine (*Pinus flexilis*) woodland knowledge, research needs, and management considerations. In The future of high-elevation, five-needle white pines in Western North America: Proceedings of the High Five Symposium. 28-30 June 2010, Missoula, MT, Proceedings RMRS-P-63, 29-36. Fort Collins, CO: US Department of Agriculture, Forest Service, Rocky Mountain Research Station.

Meko, DM, R Touchan, and KJ Anchukaitis. 2011. Seascorr: a matlab program for identifying the seasonal climate signal in an annual tree-ring time series. Computers & Geosciences 37:1234-1241. doi.org/10.1016/j.cageo.2011.01.013.

PRISM Climate Group. 2018. Oregon State University. https://prism.oregonstate.edu/explorer/. Accessed 28 July 2018.

Raffa, KF, BH Aukema, BJ Bentz, AL Carrol, JA Hicke, MG Turner, and WH Romme. 2008. Cross-scale drivers of natural disturbances prone to anthropogenic amplification: the dynamics of bark beetle eruptions. BioScience 58:501-517. doi.org/10.1641/B580607.

Resler, LM, and DF Tomback. 2008. Blister rust prevalence in krummholz whitebark pine: implications for treeline dynamics, northern Rocky Mountains, Montana, USA. Arctic, Antarctic, and Alpine Research 40:161-170. doi.org/10.1657/1523-0430(06-116)[RESLER]2.0.CO;2.

Rochner, ML, KJ Heeter KJ, GL Harley, MF Bekker, and SP Horn. 2021. Climate-induced treeline mortality during the termination of the Little Ice Age in the Greater Yellowstone Ecoregion, USA. The Holocene 31(8):1288-1303. doi.org/10.1177/09596836211011656.

Schoettle, AW, KS Burns, CM Cleaver, and JJ Connor. 2019. Proactive limber pine conservation strategy for the Greater Rocky Mountain National Park Area, General Technical Report, RMRS-GTR-379. Fort Collins, CO: US Department of Agriculture, Forest Service, Rocky Mountain Research Station. 81 pp.

Sniezko, RA, R Danchok, DP Savin, et al. 2016. Genetic resistance to white pine blister rust in limber pine (*Pinus flexilis*): major gene resistance in a northern population. Canadian Journal of Forest Research 46:1173-1178. doi.org/10.1139/cjfr-2016-0128.

Speer, JH. 2010. Fundamentals of tree-ring research. University of Arizona Press, Tucson, AZ USA.

Steele, R. 1990. *Pinus flexilis* James limber pine. Silvics of North America 1:348-354.

United States Fish and Wildlife Service. 2020. Endangered and threatened wildlife and plants; Threatened species status for *Pinus albicaulis* (whitebark pine) with section 4(d) rule. 85 Federal Register. 77408-77424.

Westerling, AL, MG Turner, EA Smithwick, WH Romme, and MG Ryan. 2011. Continued warming could transform Greater Yellowstone fire regimes by mid-21st century. Proceedings of the National Academy of Sciences 108(32):13165-13170. doi.org/10.1073/pnas.1110199108.

Windmuller-Campione, MA, and JN Long. 2016. Limber pine (*Pinus flexilis* James), a flexible generalist of forest communities in the Intermountain West. PLoS ONE 11:e0160324. doi.org/10.1371/journal.pone.0160324.

Zang, C, and F Biondi F. 2015. treeclim: an R package for the numerical calibration of proxy-climate relationships. Ecography 38:431-436. doi.org/10.1111/ecog.01335.

Subalpine Tree Species Classification Using Remote Sensing Methods and Techniques

Stephanie Jouvet, School of Environment and Geomatics, Selkirk College, BC, Canada; **Brendan Wilson**, <bwilson@selkirk.ca>, School of Environment and Geomatics, Selkirk College, BC, Canada.

ABSTRACT

Whitebark pine (*Pinus albicaulis*) can be found throughout western Canada and the United States. The species is listed as endangered federally in Canada due to comprehensive impacts such as white pine blister rust, pine beetle, fire suppression and climate change. Understanding the species' distribution and occurrence will aid in establishing localized recovery strategies to help conserve this keystone species. Remote sensing has streamlined traditional field-based data collection methods, reducing the time and resources needed to map detailed forest information over large spatial extents. The objective of this research was to explore the use of multispectral satellite imagery to distinguish between whitebark pine and two other main subalpine tree species within the Darkwoods Conservation Area in southeast British Columbia, Canada. An Object-Based Image Analysis (OBIA) and the maximum likelihood classification algorithm were used to classify WorldView-2 imagery. The mean spectral signatures of the three tree species had similar values across the eight bands available in the imagery, but Engelmann spruce has slightly lower reflectance values in the Red Edge, NIR 1 & 2 bands. The overall producer's accuracy was 68.8% and the Kappa coefficient was 64.3%. Whitebark pine's individual classification accuracy was low at 21%, compared to that of spruce (68%) and fir (63%). The overall classification accuracies were relatively low compared to other similar studies. This could have been due to several factors including a smaller sample size of geo-referenced trees, the quality of reference data points, the classification algorithm used, or that the available spectral signatures for species were too similar.

Keywords: whitebark pine, WorldView-2, satellite, Object-Based Image Analysis, species identification

INTRODUCTION

Whitebark Pine

Whitebark pine (*Pinus albicaulis*) is a species integral to high-elevation ecosystems in western North America. It is considered a keystone species for the many ecological benefits it provides including snowpack retention, flood mitigation, protection for other species, and its high-quality food source that is relied upon by many wildlife species including the Clark's Nutcracker (*Nucifraga columbiana*) (Keane et al. 2012). In Canada, whitebark pine is federally listed under the Species at Risk Act and continues to rapidly decline due to the comprehensive impacts of white pine blister rust, mountain pine beetle, fire suppression and climate change (Achuff and Wilson 2010). Extensive recovery strategies have been implemented in efforts to mitigate those impacts (Keane et al. 2017; Shepherd et al. 2018). However, being aware of the distribution of whitebark pine across the landscape is the key to understanding and mitigating the impacts this species faces (Landenburger et al. 2008).

The extent and distribution of whitebark pine occurrence on the subalpine landscape is still uncertain due to inadequate mapping and remote occurrences being undiscovered due to complex terrain (Environment and Climate Change Canada 2017). Forest inventories are imperative in understanding ecosystem diversity, forest health and species extent and provides resource managers with the information needed to make effective localized recovery strategies (Wilson and Stuart-Smith 2002; Keane et al. 2012; Environment and Climate Change Canada 2017). Field-based assessments have generally been used to conduct forest inventories but can be time consuming and costly in acquiring up-to-date data over vast areas. The integration of remote sensing in the fields of ecology, biodiversity and conservation has streamlined data collection reducing the time and resources needed for traditional field-assessments (Wang et al. 2010; Fricker et al. 2019; Nezami et al. 2020), as well as provides detailed information over large spatial extents (Immitzer et al. 2012; Maschler et al. 2018).

Remote Sensing

Remote sensing methods and techniques are integral tools utilized by forest and land managers (Fricker et al. 2019). Applications such as tree species mapping and classification have become a popular research topics as they can provide detailed ecosystem information that can be used for many different purposes. These include the monitoring of invasive species (Maschler et al. 2018), assessing biodiversity (Fassnacht et al. 2016), building wildlife habitat models (Immitzer et al. 2012), and observing fire risk as well as forest disturbance (Fricker et al. 2019).

For classifying individual tree species within complex forest environments, high spatial and spectral resolution data are preferred (Immitzer et al. 2012). Unmanned Aerial Vehicles (UAV) can host a wide variety of sensors, including high resolution Red Green and Blue (RGB), hyperspectral, multispectral or Light Detection and Ranging (LiDAR), which makes this method of real-time data acquisition flexible and task specific. Guimarães et al. (2020) reviews recent UAV applications in forestry and includes various studies related to tree species mapping and classification. Despite the many benefits of using UAVs, smaller organizations may not have access to the desired sensors or lack the resources and manpower for operating those systems.

Spaceborne satellite imagery has become more accessible with time, and can be a viable alternative to UAVs, especially in areas inaccessible to airborne platforms (Fricker et al. 2019). Traditionally, spaceborne satellite imagery was costly and could host either high spatial or high spectral sensors. For example, high spatial resolution satellite sensors, such as Quickbird (0.7m), have a limited spectral resolution of only four bands (R, G, B and NIR), while sensors such as Landsat 7 Enhanced Thematic Mapper + (ETM+) and Sentinel-2 have higher spectral resolutions but can only maintain low to medium spatial resolutions (Immitzer et al. 2012). Landenburger (2008) mapped the regional distribution of whitebark pine in the Greater Yellowstone Ecosystem using medium spatial resolution satellite imagery (Landsat Enhanced Thematic Mapper Plus). The study had highly accurate results (74.5-94.4%), which was generally unlikely when using lower spatial resolution Landsat imagery for species level classifications. Additional factors such as using remote sensing specific classification algorithms, focusing on a single tree species, the use of considerable reference data and the availability of high quality aerial imagery were said to contribute to the overall results of this study (Landenburger et al. 2008). A study conducted by Immitzer et al. (2016), reported satisfactory results when classifying two heterogenous forest stands in Germany using Sentinel-2 imagery. That study suggested that the use of multi-temporal Sentinel-2 data may further increase overall accuracies for future studies. However, it was noted that the spatial resolution of Sentinel-2 data may not be adequate for classifying heterogenous forests to an individual tree level but could potentially be more successful at broader scales. Immitzer et al. (2016) compared the results of their study to others focused on tree species mapping and classification and found that those were lower than those using a combination of high spatial and spectral data such as WorldView-2 satellite imagery.

The WorldView-2 satellite-borne sensor was launched in 2009 and hosts a very high spatial resolution panchromatic band (0.5m) and eight high spatial resolution multispectral bands (2.0m) (Satellite Imaging Corp. 2021). The multispectral data includes four basic bands (Red, Green, Blue and Near Infrared) and four distinct bands (Coastal, Yellow, Red Edge and Near Infrared 2) said to be ideal for observing vegetation characteristics (Immitzer et al. 2012). Using WorldView-2 satellite data, Immitzer et al. (2012) was successful in classifying 10 coniferous and deciduous tree species within a mid-European submontane forest with an overall accuracy of 82%. The methodology used outperformed a number of studies that were carried out utilizing remote sensing data from various sensors with different spatial and spectral resolutions conducted from 2002 to 2012, see Immitzer et al. (2012) for a detailed list of studies.

When classifying high resolution data, such as WorldView-2 satellite imagery, an Object-Based Image Analysis (OBIA) is known to outperform a pixel-based approach (Immitzer et al. 2012). An OBIA groups individual pixels into homogenous clusters that have similar spatial and spectral properties, which can be used to consider textural, geometric and neighboring characteristics during classification (Wang et al. 2010). Immitzer et al. (2012) concluded that the object-based approach outperformed the pixel-based method by approximately 10% using WorldView-2 satellite imagery.

Parametric (Maximum Likelihood) and non-parametric (Random Forest & Support Vector Machines) classification algorithms have been widely used for tree species classification (Maschler et al. 2018). Studies have compared different classification algorithms, however, the results depend heavily on the area being studied, the remote sensing and ancillary data used, as well as the experience and abilities of the analyst (Xie et al. 2019). Although non-parametric methods have become a more popular option for tree species classification within the past decade (Fassnacht et al. 2016), traditional parametric methods, such as the Maximum Likelihood, provide a simple and rapid way of classifying an image (Xie et al. 2019).

Objectives

Only a limited number of studies have explored the use of remote sensing for mapping whitebark pine (Landenburger et al. 2008; McDermid and Smith 2008). Additionally, few studies apply high spatial resolution and multispectral data to distinguish whitebark pine from surrounding tree species. Therefore, the objective of this research was to explore the potential of distinguishing whitebark pine from Engelmann spruce (*Picea engellmanii*) and subalpine fir (*Abies lasiocarpa*) using an OBIA approach. To achieve this objective, we completed the following:

1. Collected high-resolution drone imagery to validate reference data and manually delineate individual tree crowns,
2. Identified potential spectral differences among whitebark pine, Engelmann spruce and subalpine fir using spectral information derived from WorldView-2 imagery,
3. Conducted an OBIA and classify the WorldView-2 imagery, and;
4. Provided insight on the limitations and considerations based on current literature and methods utilized as well as recommendations for further research.

METHODS

Study Site

The study was conducted in the Darkwoods Conservation Area (Darkwoods) which spans 63,000 hectares of land located between Nelson and Creston (NCC: Darkwoods) (figure 1). Darkwoods is privately owned and operated by the Nature Conservancy of Canada and represents a large

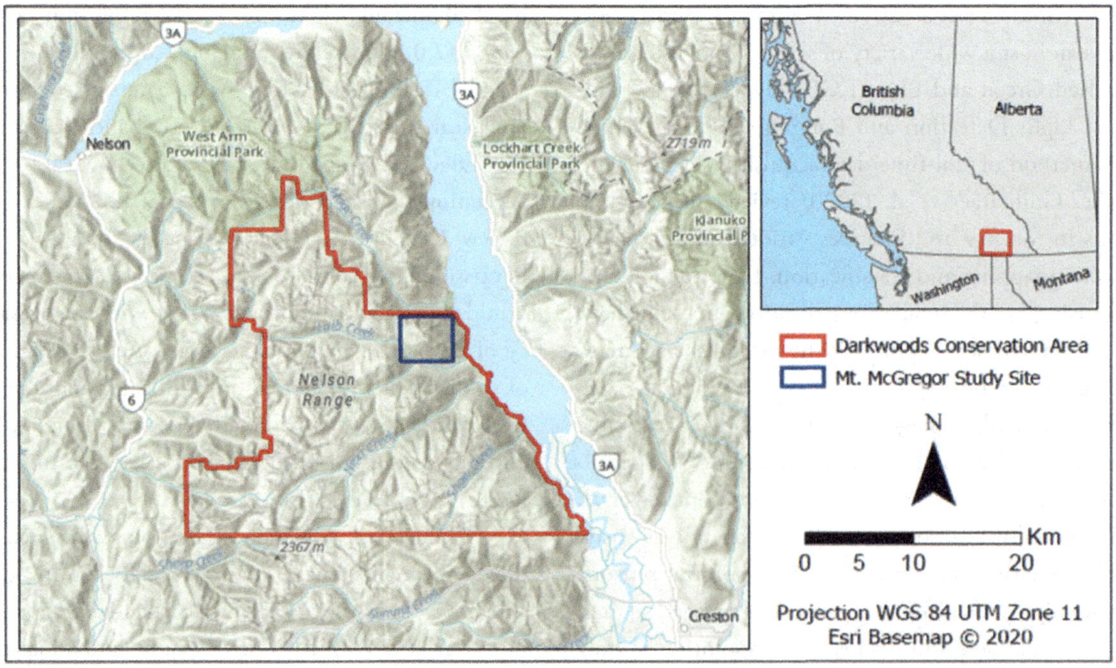

Figure 1. Darkwoods Conservation Area and McGregor site boundary, British Columbia.

tract of land used for the purpose of maintaining ecological and cultural integrity. The area consists of complex and remote terrain that encompasses valleys, mountains, lakes, creeks and diverse forests that provide important habitat to many different plant and animal species including a number of species listed as threatened or endangered by the Species At Risk Act (SARA) (NCC: Darkwoods). Field data was collected at the peak of Mount McGregor, which was identified to have whitebark pine occurrence (figure 2).

The study site lies within the Engelmann Spruce - Subalpine Fir Wet Mild Woodland (ESSFwmw) Biogeoclimatic (BEC) zone which is characterized by cold and wet summers and cold winters with a heavy persistent snowpack. This BEC zone is found in the uppermost forested areas with the elevation ranging from 1920-2150 m on cool aspects and 2000-2200 m on warm aspects (MacKillop and Ehman 2016). The dominant and co-dominant tree species in the ESSFwmw are subalpine fir and Engelmann spruce, respectively. Whitebark pine are common and generally found on drier sites. Alpine larch are also found on drier sites but are known to occupy submesic and mesic sites as well. (MacKillop and Ehman 2016).

Data

Acquisition & preprocessing of high-resolution drone imagery

The high-resolution drone imagery was recorded on August 13, 2020, using a DJI Mavic 2 Pro (figure 3). Flight conditions were clear with cloud cover and winds increasing throughout the day. The DJI Mavic 2 Pro is equipped with an RGB camera with a 1-inch CMOS sensor. The lens focal length is 28mm and has a 77-degree field of view and variable f/2.8-f/11 aperture. The imagery was captured at 20MP resolution (5472 x 3648 pixels). Ground control points were collected using a Trimble Geo7X receiver and were used to process the drone images in Agisoft Metashape and used to build a dense point cloud,

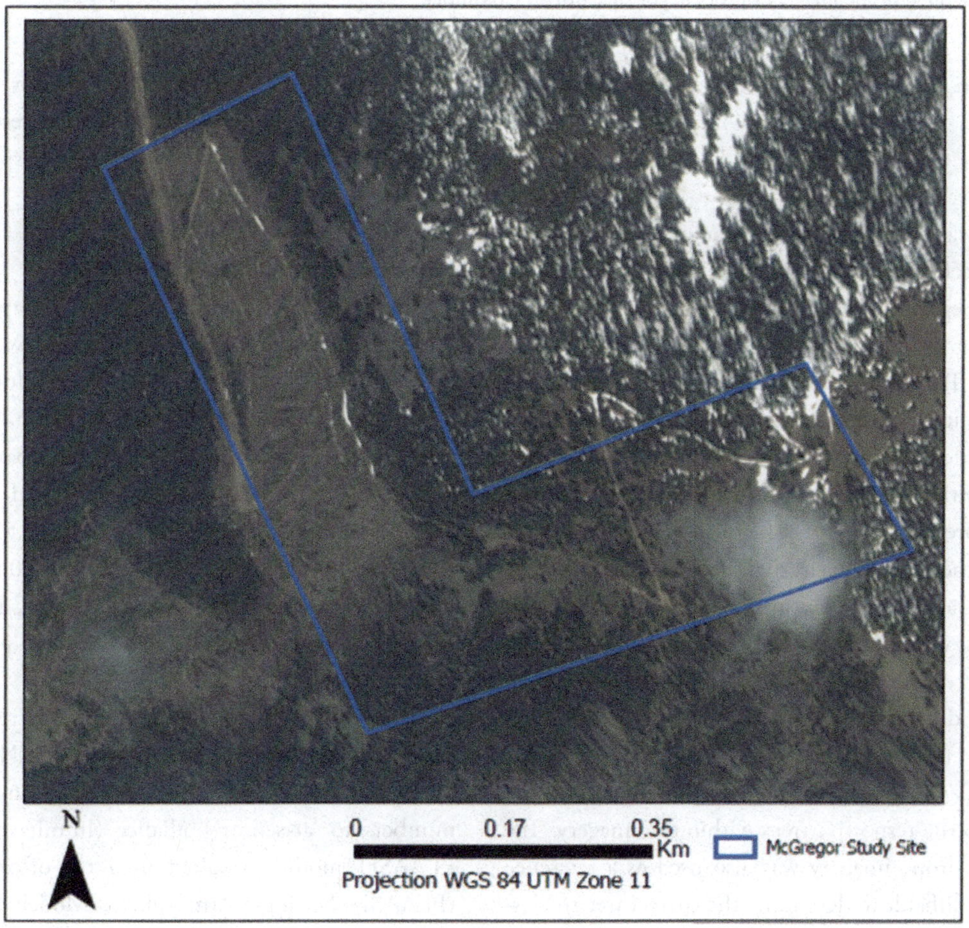

Figure 2. McGregor site located within the Darkwoods Conservation Area, British Columbia.

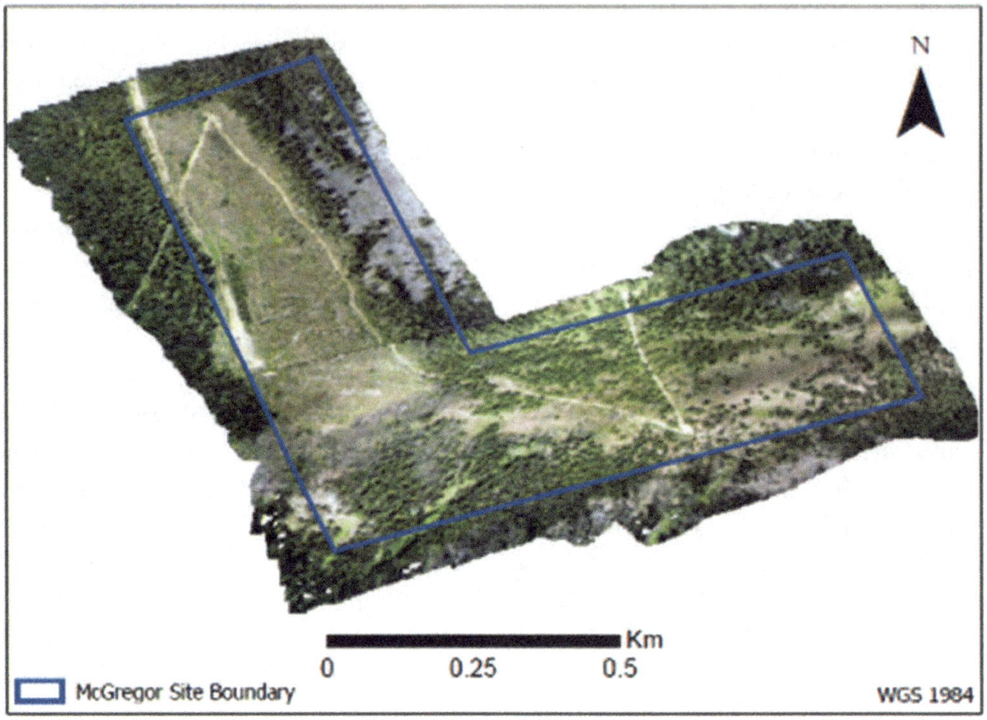

Figure 3. Orthorectified drone imagery of McGregor site, British Columbia.

Digital Elevation Model (DEM) and orthomosaic. To reduce processing time, two separate ortho images were created.

Reference data

Field data was collected over the span of 2 years with acquisition dates as follows: July 3, 2019; July 14, 2020; August 13, 2020 and September 5, 2020. Data collection consisted of geotagging trees with a high-accuracy GNSS Trimble Geo7X receiver. Tree species of interest included subalpine fir, whitebark pine and Engelmann spruce and individual trees were selected based on the intent of visually identifying the selected trees within the imagery in order to delineate the crowns. Therefore, the focus was on collecting large, isolated trees to ensure they could be clearly identified when viewing the imagery. In addition to recording the position of each tree, the species, vigor, diameter at breast heigh (DBH) and canopy area were collected. The Trimble points were post processed using differential GPS (Pathfinder Office) and exported to a shapefile format.

The WorldView-2 imagery was used to visually assign the field data to the respective trees within the imagery. The high-resolution drone imagery was also used as a reference when it proved difficult to determine the correct tree to classify within the WorldView-2 imagery. In addition, five more classes were created by visual interpretation (bare ground, cloud, shadow and vegetation). This was done referencing the WorldView-2 imagery in order to represent the other features within the image and identify any spectral differences between them and the tree classes.

Worldview-2 satellite-borne imagery

WorldView-2 satellite imagery was captured on June 2, 2019, covering the entire study area, and with minimal cloud cover (2.3%) (figure 4). The ground resolution at nadir was 30 cm and 50 cm for the panchromatic bands (0.46-0.80 µm) and 200 cm for the multispectral bands (Immitzer et al. 2012). The multispectral imagery had eight bands that includes Blue (0.45-0.51 µm), Green (0.51-0.58 µm), Red (0.63-0.69 µm), Near Infrared 1 (0.77-0.90 µm), Coastal (0.40-0.45 µm), Yellow (0.59-0.63 µm), Red Edge (0.71-0.75 µm) and Near Infrared 2 (0.86-1.04 µm) (Satellite Imagery Corp. 2021).

Radiometric and Atmospheric Corrections were applied to the multispectral imagery in ENVI 5.6 (64-bit). The Radiometric Calibration module converted the digital numbers to 'at-sensor' radiance (Immitzer, 2012) and the FLAASH module resulted in a top-of-canopy reflectance (FLAASH Settings: Atmospheric Model: Sub-Arctic Summer, Aerosol Model: Rural, Initial Visibility: 60 km), to produce meaningful spectral profiles. The multispectral im-

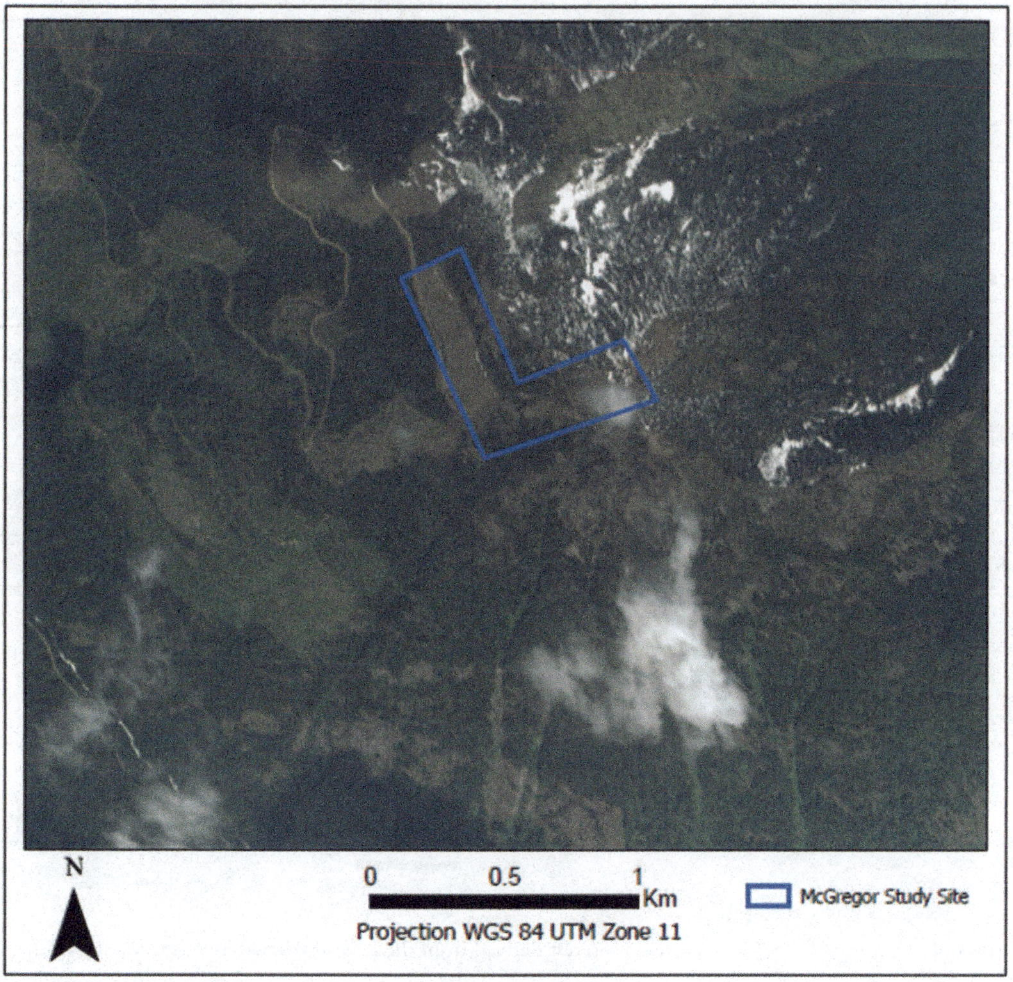

Figure 4. WorldView-2 Satellite Imagery purchased for the McGregor site.

agery was further pansharpened using the 30 cm panchromatic band and orthorectified using the Gram-Schmidt tool as well as a 25 m resolution digital elevation model (DEM)

Classification & validation

The reference data was randomly divided into two datasets, training and validation. The training data was used to conduct an example-based classification, in ENVI, that used the maximum likelihood algorithm. This algorithm assumes that the statistics for each of the classes are normally distributed and assigns the pixels with the highest probability of being within that class (Harris Geospatial Solutions, Inc. 2020). The maximum likelihood algorithm is generally biased to small sample sizes but was ultimately selected due to its ease of use and availability within the ENVI software (Fassnacht et al. 2016).

The validation dataset was used to generate fifty random points within each class. The random points were additionally assigned the classification results and used to conduct the confusion matrix to assess the accuracy of the classification.

RESULTS

Spectral Variability Among Classes

The mean spectral signatures were calculated by averaging the pixel values for all polygons within each class and are found below in figure 5. The reflectance values for snow, cloud and vegetation differ greatly from the three tree species making it easy to distinguish them from the tree species of interest. Bare shows similarities in the Blue and Red Edge bands but differ slightly in the Yellow and Red. Shadow has similarities in the Blue and Red bands but differs in the Red Edge and NIR 1 & 2 bands. Figure 6 summarizes the mean spectral signatures of the three tree species of interest. They have similar values across the bands but Engelmann spruce

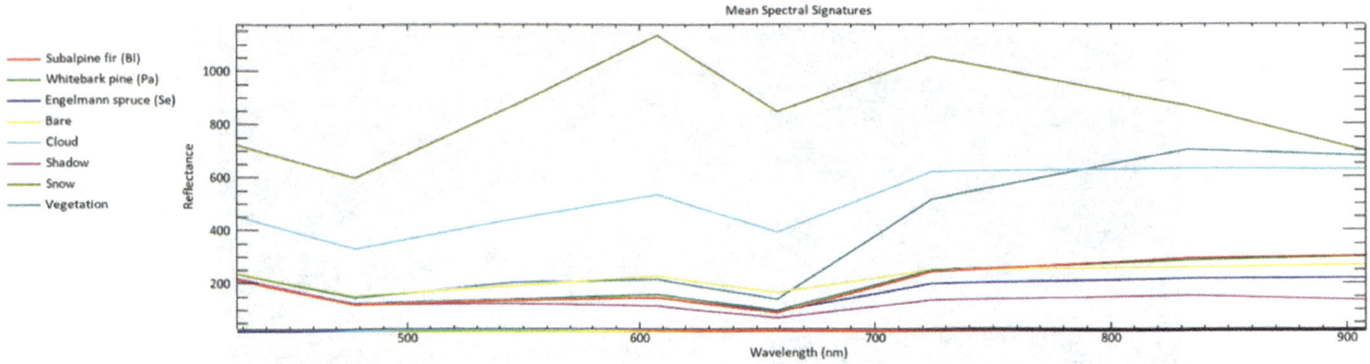

Figure 5. Mean spectral signatures of all classes derived from the 8 WorldView-2 bands using the delineated polygons.

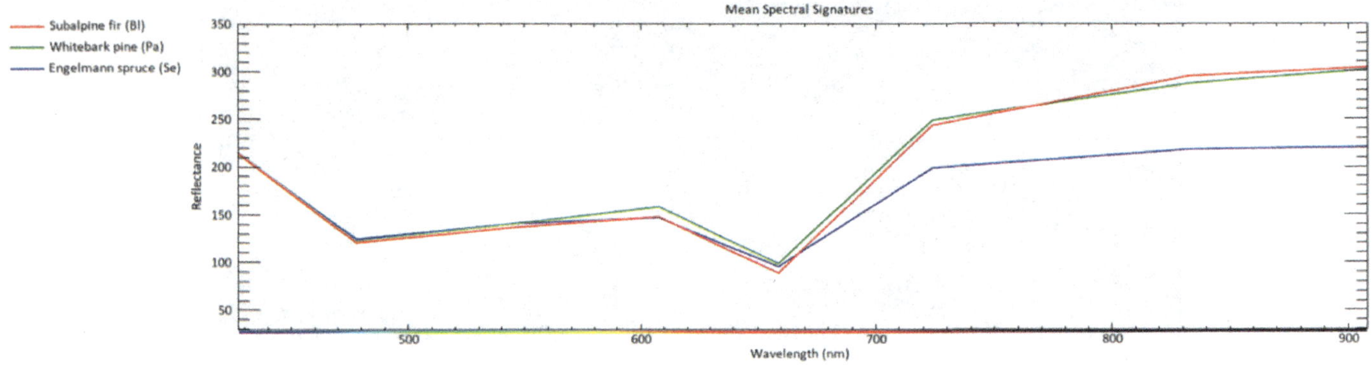

Figure 6. Mean spectral signatures of the 3 tree species of interest derived from the 8 WorldView-2 bands using the delineated polygons.

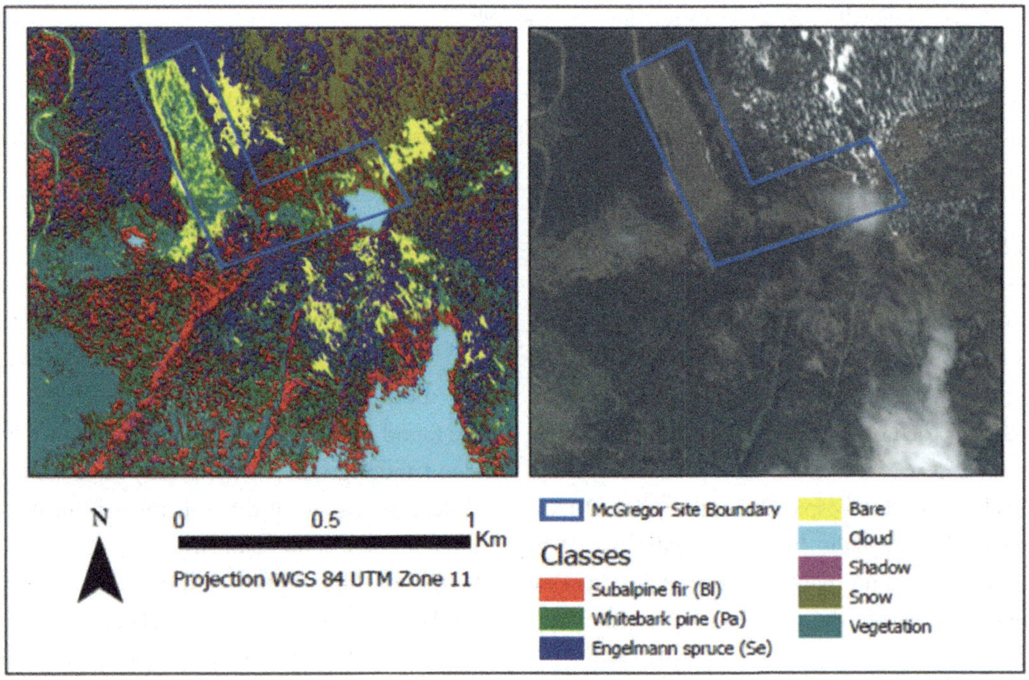

Figure 7. Classified WorldView-2 imagery.

has slightly lower reflectance values in the Red Edge, NIR 1 & 2 bands.

Validation of Classification

The confusion matrix in Table 1 outlines the accuracy results of classifying the WorldView-2 Imagery (figure 7) using all eight bands. The overall accuracy was 68.75% and the Kappa coefficient was 64.29%. The shadow and whitebark pine classes had the highest and lowest producer's and user's accuracies, respectively. Of the tree species, Engelmann spruce had the highest producer's and user's accuracies.

Table 1. Confusion matrix summarizing the classification results of the WorldView-2 Imagery using the Maximum Likelihood algorithm.

	Reference Data									
Class	Subalpine fir (Bl)	Whitebark pine (Pa)	Engelmann spruce (Se)	Bare	Cloud	Shadow	Snow	Vegetation	Total	User's Accuracy
Subalpine fir (Bl)	17	10	0	0	0	0	0	0	27	0.62963
Whitebark pine (Pa)	12	7	2	2	0	0	5	4	32	0.21875
Engelmann spruce (Se)	4	9	43	1	2	0	4	0	63	0.68254
Bare	5	7	0	37	0	0	12	1	62	0.596774
Cloud	3	0	0	0	48	0	1	0	52	0.923077
Shadow	0	0	0	0	0	50	0	0	50	1
Snow	3	3	5	3	0	0	28	0	42	0.666667
Vegetation	6	14	0	7	0	0	0	45	72	0.625
Total	50	50	50	50	50	50	50	50	400	
Prouducer's Acc.	0.34	0.14	0.86	0.74	0.96	1	0.56	0.9		0.6875
Kappa	0.642857									

DISCUSSION & RECOMMENDATIONS

Small Sample Sizes

Due to the scope of the project, there were limited field days in which data could be collected, resulting in small data sets for each of the tree species of interest. Studies have found that when conducting tree species classifications, the classes with the smallest sample sizes generally result in the lowest accuracies while larger sample sizes and extensive data sets are said to increase the success of the overall classification results for those specific classes (Landenburger et al. 2008; Immitzer et al. 2012; Grabska et al. 2019).

Data was collected for an additional site within the Darkwoods area, however, due to time constraints and difficulty confidently identifying the tree crowns with the satellite imagery, the data was not included in this study. Although the additional site data would not be significant in size, it would increase the sample size slightly and potentially result in a higher classification accuracy or encourage less spectral overlap between the whitebark pine and subalpine fir classes.

Quality of Reference Data

The study was conducted at a site with complex terrain and a highly diverse forest network which could have made it difficult to collect and validate the reference data accurately. There was an element of subjectivity when matching the geotagged points to the tree crowns as it was done based on visual interpretation of the WV-2 imagery. Isolated trees were easiest to reference, however, in heterogenous stands, it was difficult to confidently assign the points to its respective tree.

Initially, the high-resolution drone imagery was used to delineate the tree crowns. However, there were slight visual differences between the drone imagery and WV-2 imagery that caused positional inaccuracies when overlaying the polygons onto the WV-2 imagery. Therefore, the WV-2 imagery was used to visually delineate the polygons and the drone imagery was used to reference tree crowns that were difficult to distinguish. The WV-2 imagery had a lower spatial resolution than the drone imagery making it harder to confidently delineate the tree crowns and could have resulted in spectral overlap between classes reducing the overall classification accuracies for those classes (Immitzer et al. 2012).

In order to reduce the subjectivity of the manual delineation process, it is recommended that this be completed by a number of individuals, compared, and then include only the confidently identified data. This method, however, would be very time consuming and may still have a level of subjectivity imposed by the analysts. Automation tools are also available, such as the mean shift algorithm used in Maschler et al. (2018), that could remove some level of subjectivity within the segmentation step. Overall, object extraction is an essential step in achieving highly accurate results when classifying tree species, however, both manual and automated approaches are time consuming and difficult to implement properly (Michałowska and Rapiński 2021).

LiDAR

Recent studies have suggested that LiDAR has demonstrated great potential for mapping out forest environments (Michałowska and Rapiński 2021). Specifically, the ability of LiDAR technology to extract single tree parameters such as location, height, crown size, DBH and biomass estimates, by way of the three-dimensional point clouds created (Michałowska and Rapiński 2021). LiDAR derived data can be divided into three components: geometric, radiometric and full-waveform metrics that provide their own advantages for remote sensing research that are outlined in the review conducted by Michałowska and Rapiński (2021). Geometric components of generated LiDAR point clouds, for example, can provide information on tree foliage and branching structures (Michałowska and Rapiński 2021), which could be particularly useful for distinguishing between whitebark pine and subalpine fir as they are structured very differently. LiDAR in combination with other remote sensing data, including multispectral or hyperspectral imagery, has become widely used in remote sensing research today and has been found to increase the overall accuracies of tree species classification studies (Zhao et al. 2020).

Classification Algorithm

The maximum likelihood algorithm was used due to the scope of the project as well as it being easily available within ENVI (Fassnacht et al. 2016). However, this algorithm is said to be more suited to larger sample sizes and data this is normally distributed which could have contributed to the low classification results of the study (Fassnacht et al. 2016).

Traditionally, parametric classification algorithms such as supervised maximum likelihood, K-means or ISODATA have been widely used to distinguish between tree species (Fassnacht et al. 2016). However, non-parametric methods have become a great alternative as they do not require input data to be normally distributed and may be less sensitive to input variables (Fassnacht et al. 2016). Random Forest (RF) is a non-parametric method that has become popular within the world of remote sensing and has been used to successfully assess species diversity (Mallinis et al. 2020), conduct forest stand mapping (Grabska et al. 2019) and identify spectral characteristics for individual tree species classification (Immitzer et al. 2012). Immitzer et al. (2012) compared the use of Random Forest (non-parametric) and Linear Discriminant Analysis (LDA) (parametric) classifiers to distinguish between tree species in a diverse mid-European forest in Austria. The results concluded that the RF performed relatively the same as the established LDA classifier, however, RF was found to be more flexible when dealing with small sample sizes (Immitzer et al. 2012).

Should this study be continued, it is recommended that the Random Forest algorithm is used as it has been used in many tree species classification studies with high success rates (Fassnacht et al. 2016) and is ideal for dealing with small sample sizes (Immitzer et al. 2012). Despite the importance of selecting the appropriate classification algorithm for each study, collecting quality reference data may be a more important aspect to focus on to increase the overall classification results (Maschler et al. 2018).

CONCLUSION

This small-scale study evaluated the potential use of WorldView-2 imagery to distinguish whitebark pine from two separate subalpine species, subalpine fir and Engelmann spruce, at a highly diverse forest site in British Columbia. Multispectral WorldView-2 imagery was used to determine spectral differences among the tree species of interest as well as to classify the study site using the maximum likelihood algorithm. The overall classification accuracies were relatively low compared to other studies conducting tree species classification (Immitzer et al. 2012), which could have been due to a number of factors including small sample sizes, the quality of reference data, classification algorithm used, and perhaps the inherent closeness of the species spectral signatures. Should this study be continued, it is recommended to collect additional reference data to build a more extensive dataset, explore the combination of LiDAR and multispectral imagery as well as select a classification algorithm such as RF that is more flexible when working with small data sets (Immitzer et al. 2012). Although, the WorldView-2 Satellite sensor is still relatively new, there have been many recent studies that focus on tree species classification within diverse forests and complex terrain (Immitzer et al. 2012; Fassnacht et al. 2016; Immitzer et al. 2016; Fricker et al. 2019; Xie et al. 2019; Nezami et al. 2020; Michałowska and Rapiński 2021). As remote sensing technologies become more advanced and widely available, further studies should focus on distinguishing endangered species such as whitebark pine. This can streamline the process of identifying species distributions among the landscape so resource managers can implement effective large and small-scale recovery strategies to protect imperative species, such as whitebark pine, and help maintain the ecological integrity of the ecosystems they inhabit.

ACKNOWLEDGMENTS

We would like to thank Adrian Leslie and the Nature Conservancy of Canada for the opportunity and support in conducting this research within the Darkwoods Conservation Area. We are grateful to the Selkirk College students that helped with data collection and the Selkirk faculty that provided support throughout the entirety of the project. This project was supported financially by the Nature Conservancy of Canada, and with funds from the Natural Sciences and Engineering Research Council of Canada Thank-you to Dr. Garth Mowat and Adrian Leslie for providing review comments on this paper.

REFERENCES

Achuff PL, B Wilson. 2010. COSEWIC assessment and status report on the whitebark pine, *Pinus albicaulis*, in Canada. Ottawa: Committee on the Status of Endangered Wildlife in Canada. http://publications.gc.ca/site/eng/372414/publication.html. Accessed 23 September 2014.

Environment and Climate Change Canada. 2017. Recovery Strategy for the Whitebark Pine (*Pinus albicaulis*) in Canada [Proposed]. Ottawa: Environment and Climate Change Canada Strategy Series.

Fassnacht FE, H Latifi, K Stereńczak, A Modzelewska, M Lefsky, LT Waser, C Straub, A Ghosh. 2016. Review of studies on tree species classification from remotely sensed data. Remote Sensing of Environment. 186:64–87. doi:10.1016/j.rse.2016.08.013.

Fricker GA, JD Ventura, JA Wolf, MP North, FW Davis, J Franklin. 2019. A convolutional neural network classifier identifies tree species in mixed-conifer forest from hyperspectral imagery. Remote Sensing. 11(19):2326. doi:10.3390/rs11192326.

Grabska E, P Hostert, D Pflugmacher, K Ostapowicz. 2019. Forest stand species mapping using the sentinel-2 time series. Remote Sensing. 11(10):1197. doi:10.3390/rs11101197.

Guimarães N, L Pádua, P Marques, N Silva, E Peres, JJ Sousa JJ. 2020. Forestry remote sensing from unmanned aerial vehicles: a review focusing on the data, processing and potentialities. Remote Sensing. 12(6):1046. doi:10.3390/rs12061046.

Harris Geospatial Solutions, Inc. 2020. Classification tutorial. https://www.l3harrisgeospatial.com/docs/classification.html. Accessed 16 April 2021.

Immitzer M, C Atzberger, T Koukal. 2012. Tree species classification with random forest using very high spatial resolution 8-band worldview-2 satellite data. Remote Sensing. 4(9):2661–2693. doi:10.3390/rs4092661.

Immitzer M, F Vuolo, C Atzberger. 2016. First experience with sentinel-2 data for crop and tree species classifications in Central Europe. Remote Sensing. 8(3):166. doi:10.3390/rs8030166.

Keane RE, LM Holsinger, MF Mahalovich, DF Tomback. 2017. Evaluating future success of whitebark pine ecosystem restoration under climate change using simulation modeling: Simulating climate impacts on restoration treatments. Restoration Ecology. 25(2):220–233. doi:10.1111/rec.12419.

Keane, RE, DF Tomback, CA Aubry, AD Bower, EM Campbell, CL Cripps, MB Jenkins, MF Mahalovich, M Manning, ST McKinney, MP Murray, DL Perkins, DP Reinhart, C Ryan, AW Schoettle, CM Smith, 2012. A range-wide restoration strategy for whitebark pine (*Pinus albicaulis*). Ft. Collins, CO: U.S. Department of Agriculture, Forest Service, Rocky Mountain Research Station Report No.: RMRS-GTR-279. https://www.fs.usda.gov/treesearch/pubs/40884. Accessed 7 April 2019.

Landenburger L, RL Lawrence, S Podruzny, CC Schwartz. 2008. Mapping regional distribution of a single tree species: whitebark pine in the Greater Yellowstone Ecosystem. Sensors. 8(8):4983–4994. doi:10.3390/s8084983.

MacKillop, and DA Ehman. 2016. A field guide to ecosystem classification and identification for Southeast British Columbia: the South-Central Columbia Mountains. Victoria, B.C: Province of British Columbia, Ministry of Forests (Land management handbook).

Mallinis G, I Chrysafis, G Korakis, E Pana, and AP Kyriazopoulos. 2020. A random forest modelling procedure for a multi-sensor assessment of tree species diversity. Remote Sensing. 12(7):1210. doi:10.3390/rs12071210.

Maschler J, C Atzberger, and M Immitzer. 2018. Individual tree crown segmentation and classification of 13 tree species using airborne hyperspectral data. Remote Sensing. 10(8):1218. doi:10.3390/rs10081218.

Michałowska M, and J Rapiński. 2021. A review of tree species classification based on airborne lidar data and applied classifiers. Remote Sensing. 13(3):353. doi:10.3390/rs13030353.

NCC: Darkwoods. http://www.natureconservancy.ca/en/where-we-work/british-columbia/featured-projects/darkwoods/. Accessed 8 December 2019.

Nezami S, E Khoramshahi, O Nevalainen, I Pölönen, and E Honkavaara. 2020. Tree species classification of drone hyperspectral and rgb imagery with deep learning convolutional neural networks. Remote Sensing. 12(7):1070. doi:10.3390/rs12071070.

Satellite Imaging Corp. WorldView-2 Satellite Sensor | Satellite Imaging Corp. https://www.satimagingcorp.com/satellite-sensors/worldview-2/. Accessed 29 April 2021.

Shepherd B, B Jones, R Sissons, J Cochrane, J Park, CM Smith, and N Stafl. 2018. Ten years of monitoring illustrates a cascade of effects of white pine blister rust and focuses whitebark pine restoration in the Canadian Rocky and Columbia Mountains. Forests. 9(3):138. doi:10.3390/f9030138.

Wang K, SE Franklin, X Guo, and M Cattet. 2010. Remote Sensing of Ecology, biodiversity and conservation: a review from the perspective of remote sensing specialists. Sensors. 10(11):9647–9667. doi:10.3390/s101109647.

Wilson B, and J Stuart-Smith. 2002. Whitebark pine conservation for the Canadian Rocky Mountain National Parks. Winlaw, BC: Cordilleran Ecological Research Report No.: KNP01- 01.

Xie Z, Y Chen, D Lu, G Li, and E Chen. 2019. Classification of land cover, forest, and tree species classes with ziyuan-3 multispectral and stereo data. Remote Sensing. 11(2):164. doi:10.3390/rs11020164.

Zhao D, Y Pang, L Liu, and Z Li. 2020. Individual tree classification using airborne lidar and hyperspectral data in a natural mixed forest of Northeast China. Forests. 11(3):303. doi:10.3390/f11030303.

Mountain Marathon: Endangered Limber and Whitebark Pine Recovery in Alberta

Jodie Krakowski, <jodie@whitebarkpine.ca>, Whitebark Pine Ecosystem Foundation of Canada; **Robin Gutsell**, <robin.gutsell@gov.ab.ca>, Alberta Environment and Parks, Species at Risk Stewardship.

ABSTRACT

Work to recover endangered whitebark and limber pine in Alberta has been underway for over two decades. A summary of the provincial recovery and restoration program is described here. The species' slow growth, remote habitats, irregular cone crops, and non-commercial status pose unique challenges. Their large, nutrient-rich seeds are a key source of food for wildlife species in these habitats, supporting a unique relationship with Clark's nutcracker. Their persistence in extreme sites helps anchor fragile soils, sustain hydrological function in montane headwaters that support endangered salmonid populations, and initiate treeline formation. Mitigating the main threats causing their decline requires a multi-pronged, sustained effort focused on disease resistance genetics at multiple scales. Identifying and testing a genetically diverse base of well-adapted, disease-resistant trees to provide seed for natural and artificial regeneration is the core of the Alberta recovery program. This is complemented by landscape-level strategies to reduce threats caused by mountain pine beetle pressure and wildfire risk. Promoting knowledge of the value of and threats to these species raises awareness and helps avert and mitigate direct human impacts. The ranges of whitebark and limber pine cross jurisdictions and require active partnerships across borders to make recovery a success.

INTRODUCTION

Whitebark (*Pinus albicaulis*) and limber pine (*Pinus flexilis*) trees are Endangered in Alberta as a result of rapid population decline caused by shared threats. In B.C., both species are Blue-Listed (threatened). Whitebark pine is Endangered federally in Canada and limber pine is pending a federal listing decision after the Committee on the Status of Endangered Wildlife in Canada (COSEWIC) assessed it as Endangered. These keystone species provide many unique and valuable ecosystem services, the most important provided by their uniquely large and rich seeds as a food source to numerous wildlife species (COSEWIC 2010, 2014). Other important functions include moderating headwater flows in bull trout (*Salvelinus confluentus*) and cutthroat trout (*Oncorhynchus clarkii*) habitat, anchoring steep and fragile soils, catalyzing upper treeline development, and supporting important cultural values for many Indigenous nations, and place-based experiences for people who recreate and support livelihoods in these areas.

The main threats to the species in Alberta are the introduced fungal pathogen causing the fatal disease white pine blister rust (*Cronartium ribicola*) and mountain pine beetle (*Dendroctonus ponderosae*). Ancillary threats interacting with them, and with each other, include changes to wildland fire regimes leading to successional replacement by more shade-tolerant competitors, mortality caused by larger, more extreme fires related to fire suppression and climate change, and the various detrimental climate change impacts on high-elevation species. Recovery plans contain actions to mitigate each threat and support the goals of genetically diverse, self-sustaining populations of these species throughout their ranges. Whitebark and limber pine grow very slowly, adapted to their severe habitats, so recovery requires a commitment on the order of a century: a single tree generation. The remote areas they grow in pose unique logistical challenges and add high access costs. Sharing resources, knowledge and data across jurisdictions has enabled agencies to enhance capacity and achieve what no single jurisdiction or agency could. The Alberta whitebark and limber pine recovery program is outlined here, including achievements to date.

RECOVERY PLANS

Federal Status and Plans

Both species share biological and ecological characteristics, are impacted by the same threats (with some regional differences), and therefore have similar recovery plan goals and actions. Whitebark pine was assessed as Endangered in 2010 by the Committee on the Status of Endangered Wildlife in Canada (COSEWIC 2010) and listed on Schedule 1 of the *Species At Risk Act* (SARA) in 2012. The draft federal recovery plan (Environment Canada 2017) is still pending finalization. Limber pine was assessed as Endangered by COSEWIC in 2014, and is still pending a SARA listing decision, which will trigger development of a federal recovery plan.

Implementation is coordinated by a working group of technical, management, and communications specialists from agencies and jurisdictions where whitebark and limber pine grow. Participants represent Parks Canada Agency, departments of the provincial Alberta and British Columbia (B.C.) governments, and the Whitebark Pine Ecosystem Foundation of Canada. A series of facilitated Conservation Standards (formerly Open Standards) workshops convened by Parks Canada Agency and now hosted by the Whitebark Pine Ecosystem Foundation of Canada began in 2018. This process has resulted in the development of recovery goals, objectives, indicators, and pathways for action.

Provincial Status and Plans

In 2008, Alberta listed whitebark and limber pine as *Endangered* under the *Wildlife Act*. A provincial recovery team included members from provincial and federal government agencies and non-government organizations, with stakeholder meetings including industry representatives. The responsible Minister approved recovery plans developed by the team for each species (Alberta Whitebark and Limber Pine Recovery Team 2014a,b). Implementation then became the focus, with recovery team membership consolidated to provincial government. Co-chairs represented Alberta Environment and Parks and Alberta Agriculture and Forestry, supplemented by team members with diverse expertise. The recovery plan was completely revised in 2019 (pending approval) to combine both species and integrate streamlined Open Standards objectives and targets. In 2020, Alberta Agriculture and Forestry reduced their involvement and the Whitebark Pine Ecosystem Foundation of Canada became co-chair; at the same time, external experts and agency representatives were added to the team.

Related plans support specific aspects of recovery. Both in situ and ex situ gene conservation plans for the province provide snapshots of the status for native Alberta tree species and identify conservation gaps and priorities (Alberta Agriculture and Forestry 2018a,b). Recovery and restoration plans have been developed throughout the species' ranges, including a range-wide U.S. strategy supported by the Whitebark Pine Ecosystem Foundation (Keane et al. 2017; Tomback et al. 2022). Restoration work in U.S. National Forests (Jenkins et al. in press), National Parks, the Greater Yellowstone area, Bureau of Land Management lands (Perkins et al. 2016), and some Tribal Lands is guided by whitebark and limber pine restoration plans. B.C. has plans in place for several regions and provincial parks (Clason 2013, Wilson and Stuart-Smith 2001), and the province developed an implementation strategy [in review, K. Bennett, BCMFLNRORD, personal communication]. The Crown Managers Partnership High Five Working Group developed a spatially explicit whitebark and limber pine restoration strategy covering the Crown of the Continent Ecosystem (Jenkins et al. in press) that Alberta will be implementing. Parks Canada has incorporated these species in their Multi-Species Action Plans (Parks Canada Agency 2017a,b,c); and there are numerous mitigation plans and best practices for specific activities and regions to reduce direct impacts to and improve outcomes for these species.

The long-term provincial recovery plan goal is: "to have at least one self-sustaining metapopulation per species and per management unit, of sufficient size, composition, and distribution to sustain Clark's nutcracker populations within the historical range of whitebark and limber pine and support adaptation in their projected future range." Supporting objectives are based on regional threats and impacts. Strategies are outlined to meet objectives, with progress measures and actions required for each objective (table 1). The slow growth and maturity of these trees, which do not begin producing cones until age 50 (limber pine) to 80 (whitebark pine) years means that objectives ultimately aim a century down the road, to gauge effectiveness of recovery actions now.

Active support from collaborators within and across jurisdictions has been integral to the Alberta recovery program. The diversity and sharing of perspectives and resources has been invaluable to standardize methodology, improve data col-

Table 1. Goal, objectives, and strategy of Alberta's whitebark and limber pine recovery plan.

Goal	Objective	Strategy
To have at least one self-sustaining metapopulation per species and per management unit, of sufficient size, composition, and distribution to sustain Clark's nutcracker populations within the historical range of whitebark and limber pine and support adaptation in their projected future range	1. By 2100, the rate of increase in the metapopulation of five-needle pine trees with elevated disease tolerance or resistance in each management unit is greater than the rate of decline caused by blister rust 2. By 2120, at least one self-sustaining metapopulation of each five-needle pine species is established north and south of Highway 1.	1. Maximize the frequency of disease-resistant trees in five-needle pine habitat in order to reverse the decline caused by white pine blister rust, supported by: 1.1. Identify, protect and test plus trees (i.e., trees selected in the field for disease resistance). 1.2. Develop at least one seed orchard for each species sufficient to supply seed with increased disease resistance to meet restoration needs. 1.3: Restore populations in suitable habitat to sustain ecological function. 2. Restore fire regime in five-needle pine habitat within the historical range of variability. 3. Address priority knowledge gaps.

lection, share information, leverage resources, fill knowledge gaps, access sites and materials, and build support. A list of key partners by category is below.

- B.C.: Ministry of Forests, Lands and Natural Resource Operations and Rural Development, B.C. Parks
- Canada: Parks Canada Agency, Natural Resources Canada—Canadian Forest Service Pacific, Northern, Laurentian, and Atlantic Forestry Centres
- U.S.: U.S. Department of Agriculture, Forest Service: Dorena Genetic Resource Center, Coeur D'Alene Forest Nursery, Rocky Mountain Research Station, Intermountain Research Station
- Academia: The King's University (Edmonton), Montana State University, University of Alberta, University of British Columbia, University of Calgary, University of Northern British Columbia, University of Victoria
- Non-governmental organizations: Whitebark Pine Ecosystem Foundation of Canada, Crown Managers Partnership High Five Working Group, Nature Conservancy of Canada
- Many private landowners whose property contains limber pine stands and plus trees (defined below)

RECOVERY ACCOMPLISHMENTS

Short-term and long-term actions must occur simultaneously. Sequential actions are guided by the Conservation Standards "results chains". Specific actions appropriate to each area or suite of habitat characteristics depend on the condition and status of the site, stand or population. Progress in each area is summarized below.

Improved Spatial Habitat Models

Development impacts to these species can be avoided or more effectively mitigated with accurate spatial data, such as timber harvest, linear project or ski area establishment. Provincial forest inventory (Alberta Vegetation Inventory) has poor accuracy for whitebark and limber pine, as design and sampling focus on commercial forest types. Where the two species overlap, site visits are required to confirm identification due to their similar appearance.

Spatial habitat suitability models were developed adapting McDermid and Smith (2008) for limber and whitebark pine in Alberta (excluding National Parks which had a similar project underway). Digital elevation models (DEM) at 25 m resolution were used in the few townships lacking LiDAR at 1 m resolution. A randomized training data set of presence and absence points and polygons was used to build the model, and one third of the data were reserved for validation. High and moderate whitebark pine habitat suitability classes and one suitable class for limber pine were supported (table 2a,b).

Spatial habitat suitability models (LiDAR and DEM for each species) are available for public download through Alberta's Open Data web portal. While not perfect, the resulting models were a significant improvement over previous range maps and accuracy assessments by township support their use for interpretation at the stand or population scale, and not to predict the location of individual trees. Stochastic factors such as competition, soils, disturbance, and bird-mediated dispersal caused discrepancies between the actual and predicted niches. A due-diligence check based on available records, imagery and a field check is recommended before proceeding with plans or activities.

These models, as well as field location records, are incorporated in updated spatial layers for the Government of Alberta Wildlife Sensitivity Layers, which support the Landscape Analysis Tool that permit applicants are required to use to determine potential overlaps with species at risk habitat, dispositions, and other features that may require design and mitigation measures.

Density models were explored, assigning density surfaces to the habitat model to infer stem counts based on average tree crown dimensions, but results were not accurate enough to use. As more field data is collected, density models may be pursued in the future.

Long-Term Monitoring

A network of over 250 long-term monitoring plots is established throughout the Rocky Mountains of Alberta and B.C. Each whitebark and limber pine tree taller than 1.4 m is tagged and reassessed for health, growth, form, and damage. Living regeneration is tallied by health status. These data are critical to quantify growth, regeneration, health, and mortality status and trends for national and regional

Table 2a. Modelling parameters for whitebark pine habitat probability in Alberta. Columns indicate parameter combinations and values applicable to each suitability category.

Parameter	Values	Low (exclude)	Low (exclude)	Low (exclude)	Moderate	High	High	High
Slope	50-150%			+		+	+	+
Elevation	1750-2250 m		+	+	+	+	+	+
Aspect	112.5-270°				+	+		+
Topography	Ridge						+	+
Canopy height[a]	3-30 m	+	+	+	+	+	+	+

[a]LiDAR model only

Table 2b. Modelling parameters for limber pine habitat probability in Alberta. Columns indicate parameter combinations and values applicable for suitable habitat.

Parameter	Values	Suitable	Suitable	Suitable	Suitable	Suitable
Slope	30-150%			+	+	+
Elevation	1450-1800 m	+	+	+	+	+
Elevation	1450-1800 m		+	+	+	+
Aspect	212-270°		+	+		+
Topography	Ridge	+	+		+	+
Canopy height[a]	3-30 m	+	+	+	+	+

[a]LiDAR model only

reporting. Stand- and region-specific recovery actions are guided by the monitoring results. These plots help identify areas where infection is high to target plus tree selection. The data can also be used to identify areas where natural regeneration is lagging mortality rates and requires restoration planting. Trends over time highlight regional and ecological pressures, such as rate of increase in severity and extent of blister rust and mountain pine beetle, as well as abiotic stressors (Shepherd et al. 2018; Smith et al. 2008, 2013a,b). These plots are identified provincially as high-value natural resources for wildfire protection, and to incorporate in planning and permitting.

Plots were established in the mid-1990s, with agencies gradually adding more. Starting in 2003-04, Parks Canada and Alberta worked to standardize all plots following published methodology by the Whitebark Pine Ecosystem Foundation (Tomback et al. 2005). The network of plots is remeasured every five years. Staff from different agencies now train together, co-ordinate planning, and share resources to ensure efficient and consistent data collection. Many sites require long hikes, helicopter access, or landowner permission to access. New plots are established to fill spatial and ecological gaps, and where mortality is extreme and access is lost, some plots are dropped (table 3).

Table 3. Current status of long-term monitoring plots.

Plots	Whitebark pine	Limber pine	Total
2019 assessed	80	164	244
Total plots	95	205	300

Plus Trees

Plus trees are trees that appear phenotypically disease-resistant in the field based on standardized assessment methodology (long-term monitoring plots or 100-tree survey temporary plots), but have not yet been tested to confirm heritable resistance. Priority stands have blister rust infecting at least 75%, ideally over 85% of pines. Trees exhibiting signs of resistance, or that are substantially healthier than the surrounding stand, that may not be symptom-free, are plus tree candidates (Mahalovich and Dickerson 2004). Field teams may assess hundreds of trees before selecting one plus tree. To ensure genetic diversity is maximized by not sampling from related groups of trees, plus trees cannot be less than 50 m apart.

The entire crown and stem of each plus tree is carefully evaluated for signs of blister rust with binoculars, ideally by two observers. Trees should be reproductively mature so seed can be collected for resistance testing. Scions (branch tip cuttings from upper branches) can be collected and grafted onto rootstock to develop a genetic archive with copies of the tree. Each plus tree receives a permanent metal tag with a unique ID, and is geolocated, measured and photographed. All information is stored in the provincial recovery program database and spatial layers, updated annually, and shared with land managers as these trees represent valuable and irreplaceable genetic resources. Plus trees determined to be at elevated risk of beetle attack annually are identified for protection by applying verbenone and green-leaf volatiles. This protection must be applied before beetles overwintering in the bark emerge, typically mid- to late June in these habitats.

Collections of seed before 2015 were based primarily on accessibility and area representation, yielding a broad sample of genotypes for conservation that were not selected for

Table 4. Plus trees selected for the Alberta recovery program.

Year	Selected limber pine	Selected whitebark pine	Monitored limber pine	Monitored whitebark pine
Pre-2015	0	0	0	0
2015	84	0	0	0
2016	51	47	0	0
2017	82	12	0	0
2018	2	0	163	24
2019	0	0	0	0
2020	0	0	145	28
2021	0	0	131	40

health. While 138 trees could not be confidently relocated, 383 were, but few met plus tree criteria. From 2015 onwards, field surveys focused on finding new plus trees and monitoring those in the program (table 4). Not all plus trees have been submitted for testing yet, or grafted in clone banks or seed orchards because of limited cone collection opportunities and funding. Because screening takes 7 years from when seed is submitted, no elite trees are identified yet, and plus trees are the source of all restoration seedlings.

Protection of Plus Trees from Mountain Pine Beetle

Annual overwinter survival surveys by Alberta forest health staff are used to identify areas with elevated mountain pine beetle pressure overlapping whitebark and limber pine habitat. High-value program components including plus trees are selected for protection from mountain pine beetle attack. A combination of verbenone and green-leaf volatiles in flaked packets is applied before beetle emergence, typically mid- to late-June. Studies have shown this treatment's high efficacy for whitebark pine (Cardinal et al. 2021), and it appears to be effective for limber pine as no treated plus trees have suffered fatal attacks to date. Where feasible, this is combined with other operational field work for efficiency.

Seed Collections

Seed collections are archived for long-term genetic conservation; used for scientific research; grown into seedlings for operational restoration planting; and exchanged with other agencies to support mutual goals. The seed inventory of the province is adjusted as seeds are withdrawn for use, and added to following collection and processing. Each tree, and each collection year, are tracked individually as the genetics differ because differing pollen contributions affects the genetics of the seedlot.

Collecting seed from these species is logistically challenging and very costly. Limber and whitebark pine produce moderate to heavy cone crops irregularly every three to five years (masting), with little intermittent seed. Cone crops can be assessed the previous year when they are immature, which helps with planning and securing resources. Seed needs to be collected as late as possible for the highest viability. To prevent wildlife predation from plus trees, protective wire mesh cages must be installed over cones early in the season, and removed in late fall. Cones grow only at branch tips, and each tree must be accessed and climbed twice. Cones require careful handling and time-consuming manual processing to extract and clean seeds, especially if there is a need to keep single tree seed separate, for example to test for disease resistance. Documenting properties for each parent tree and seed accession is key to tracking collections in the provincial archive at the Alberta Tree Improvement and Seed Centre in Smoky Lake.

Mast years occurred in 2010, 2015, and 2018 in Alberta, with a moderate cone year for limber pine in 2021. Collections before 2015 targeted geographic representation and accessibility, and included some collections where different trees were bulked (mixed) together (table 5).

Disease Resistance Screening

After the parent tree is selected, seeds are collected to produce seedlings that are then infected and screened to quantify the heritability of each parent tree's rust resistance, based on the performance of their offspring. Depending on the facility, the proportion and sometimes the type of resistance to blister rust that they pass on to their offspring can be characterized. Many plus trees lack heritable resistance, emphasizing the critical importance of testing to avoid continuing to use those trees for restoration, as their seedlings are likely to die from infection before they reproduce. Partly because not all plus trees have resistance, and partly because much of the resistance identified is only partial, many plus trees must be selected to deliver long-term resistance targets. Some plus trees with elevated partial resistance will still get infected and die, and only a proportion of seedlings from partially resistant parents will survive. No whitebark pine and only a small fraction of selected limber pine parent trees have complete resistance, conferred by a single dominant gene (Sniezko et al., in press). While highly valuable, this single gene is likely to be overcome by the pathogen evolving virulence, so layering multiple types of resistance is the strategy that will lead to the most widespread, durable resistance, rather than deploying the same single-gene resistance as broadly as possible. The baseline wild population level of blister rust resistance is zero (fully susceptible). Trees with a certain threshold of heritable resistance after testing are "elite" trees.

Screening takes seven years from seed (once stands are surveyed, plus trees identified, and a year with a collectible cone crop occurs). A two-year-old seedling has enough foliage to inoculate for reliable results. A garden of inoculum sources on the main alternate blister rust host (Ribes spp.) is maintained and from 25 to 144 seedlings per plus tree (varying with each program) are randomly assigned to blocks. When spores are at the appropriate stage to infect

Table 5. Seed collections at Alberta Tree Improvement and Seed Centre.

Collection year	Whitebark pine number	Whitebark pine kg	Limber pine number	Limber pine kg
Pre-2015	391,224	35.200	904,485	70.677
2015	12,706	1.626	135,208	10.779
2016	61,954	6.720	907	0.078
2017	1,598	0.273	8,365	0.773
2018	62,067	7.506	208,779	14.305
2019	0	0.000	0	0.000
2020	0	0.000	9,967	0.821
2021	18,208	2.276	72,672	6.056

pine needles, seedlings are inoculated under carefully controlled conditions so they are fully infected at high concentrations, but not overwhelmed with unrealistic concentrations, of spores. They are then monitored over four years as the different resistance mechanisms develop over time, ultimately yielding a score for the parent tree.

Several facilities have developed blister rust resistance screening capacity. The U.S. Forest Service has dedicated facilities at Coeur D'Alene Forest Nursery in Idaho (CDA) and at the Dorena Genetic Resource Centre in Cottage Grove, Oregon (DGRC). The B.C. Ministry of Forests, Lands and Natural Resource Operations and Rural Development has developed Canada's first screening facility at Kalamalka Forestry Centre in Vernon (KFC). Table 6 shows the material Alberta has sent for screening. Occasionally trees needed repeat screening as a result of various factors including poor seedling production or low inoculum concentrations, so the numbers of trees may total more than the numbers of plus trees screened. Agency collaborators have also screened parent trees for other recovery partners, such as National Parks and B.C. Parks, which are not included in these tallies, but would contribute to the material included in seed orchards if they meet the guidelines for planting in Alberta.

Table 6. Trees sent for disease-resistance screening, by sowing year.

Year	Whitebark pine	Limber pine
2010	10* CDA	0
2015	6 KFC	50 DGRC
2016	8 KFC	50 DGRC
2017	7 FKC	50 DGRC
2018	9 KFC	12 DGRC
2019	10 KFC	6 KFC
2020	0	0
2021	0	0

Restoration Planting

Planting seedlings is what restoration typically brings to mind, which represents the culmination of many of the prior and subsequent steps described here. Years of advance planning are needed to grow durable seedlings and secure authorization to plant in a suitable site. These valuable seeds must be stratified with a series of controlled moisture and temperature treatments for five to six months to ensure all potentially viable seeds germinate consistently. A two-year-old whitebark pine seedling with robust roots has the optimal balance between performance, cost, and time compared to younger seedlings, or seeds, which have very high predation rates. Working with nurseries is important to build capacity and improve results over time to grow whitebark and limber pine seedlings at an operational scale. A multi-year learning curve is needed compared to growing commercial reforestation seedlings. These significant costs mean that Alberta is focused on propagating plus tree seeds to maximize long-term survival to achieve recovery goals.

Regulatory requirements in Alberta specify a seedlot must be registered with the province before it can be planted on Crown land. All limber and whitebark pine seedlots have been registered, and have dispositions established identifying restoration projects on provincial land.

Tens of thousands of seedlings need to be grown and planted annually for significant progress in recovery. By building a program over time, planting numbers have shown an increasing trend (table 7).

Table 7. Plus tree and elite (screened) seedlings planted and hectares restored in Alberta.

Year	Whitebark pine seedlings	Whitebark pine hectares	Limber pine seedlings	Limber pine hectares
2018	1700	6.80	550[a]	5.50
2019	0	0.00	7227	36.00
2020	4400	13.28	1600	10.21
2021	200[b]	0.50	6976	12.23

[a]another 550 were planted in Waterton Lakes National Park
[b]planted in Waterton Lakes National Park

Seed Zones and Provenance Trials

To ensure that seedlings are suited to the environment where they will grow for the next century or more, their ranges in Alberta are divided into species-specific seed zones reflecting adaptation and regional differentiation. Based on long-term provenance trial results, the zones and transfer limits may be adjusted over time. Provenance trials, also called common garden studies, are long-term tests that compare performance of genotypes sourced from different areas when growing in a common environment. Provenance trials quantify how far seed can be moved (geographically or climatically) from its origin location before it becomes maladapted. A genetically diverse and well-adapted population is needed for restoration in each region. This may be challenging in regions where rust infection levels are too low to select plus trees, and alternate strategies may be more appropriate.

Alberta has one limber pine provenance trial, planted in 2016 with three-year-old seedlings that tests 145 seed sources from 30 populations sampled along the Rocky Mountains. A paired test site is at Fort Collins, Colorado, comprising the International Limber Pine Provenance Study (ILPPS) in partnership with the U.S. Forest Service Southwest Research Station and the University of B.C. Faculty of Botany. Age 5 measurements (planted; age 8 from seed) for growth and vigour were completed in 2021. Alberta also has one test site of a series of 12 whitebark pine provenance trials, plus some smaller ancillary tests that were established by the B.C. Ministry of Forests, Lands and Natural Resource Operations and Rural Development in 2015 and 2017. The Alberta site was planted with two-year-old seedlings in 2017 with 52 range-wide sources.

Seed Orchards

A seed orchard is a plantation containing copies (clonal orchard) or progeny (seedling orchard) of selected trees that is designed to maximize seed production in a single location. Seed orchards composed of the best available material will expedite whitebark and limber pine recovery. Grafted orchards are being established containing copies of the best selected or tested trees, which will be pollinated by a diverse group of the other best trees for each seed zone. As test results become available, poor performers can be removed (rogued) and new plus tree selections, or more copies of tested elite trees, can be added. Multiple seed orchards can be co-located in a single area to facilitate administration and management, especially for these species that appear to have broad transferability. A seed orchard can be outside the seed zone and ideally far from potentially contaminating pollen sources, as long as the location supports good pollen and seed production. Because these are the first seed orchards of these species and our knowledge of their reproductive biology off-site is lacking, partners agree the best approach is to establish multiple sites in several different environments.

Following a workshop among partners, a range of key topics were discussed and several have been finalized, including candidate sites, delineation of seed zones, partner contributions, and policy options and alignment to yield outcomes consistent with provincial regulatory requirements for seed orchards and seed transfer. Because Alberta contains around 90% of the Canadian range of limber pine, one orchard is established in Waterton Lakes National Park with a second candidate site planned for establishment trials in 2022 on property owned by the Calgary Zoological Society. At least two seed orchards are established for whitebark pine in B.C. The lifespan of these orchards is expected to be at least 50 years, so the right size, design, location, infrastructure, management input, and access decisions are very important.

Clone banks are key complements to seed orchards as ex situ living genetic archives. They contain copies of selected material as a backup in case the original parent tree is killed or can no longer be accessed. Clone banks can also be used for tree breeding as they mature. There are at least two whitebark pine clone banks planned or established in B.C., and plans are being developed for limber pine clone banks.

Knowledge gaps around the reproductive biology of whitebark and limber pine are being addressed through a literature review, and establishing various pilot studies including cone induction methods, different site ecology for multiple orchards, and orchard management options that will include irrigation, fertilization, and pollen management.

Habitat Thresholds

Studies have indicated that a threshold of cone-producing whitebark pine is needed to sustain Clark's nutcracker visitation (Barringer et al. 2012; McKinney et al. 2009). Because it is impractical to count cones or infer cone density over broad areas, substitutes such as stem counts or basal area per hectare of mature trees can be used. After piloting several options, Alberta has been collecting basal area with variable radius prism plots. This data can help characterize stands that are likely producing sufficient cones to consistently reproduce at the upper bound of basal area, and those that are marginal or need planting at the lower bound of basal area.

A key data gap, however, is whether these thresholds, based on stands in the Yellowstone region, correspond with ecosystems in Canada where key factors such as stand composition and Clark's nutcracker alternate food source availability differ substantially. Systematic bird telemetry studies in Canadian habitats are needed to fill this gap.

Other Restoration Projects

Optimizing seed handling and germination of whitebark and limber pine

Consistent with what other practitioners have observed, Robb (2020) found that collecting cones later promotes embryo and seed development. This effectively mimics nature: Clark's nutcrackers cache seeds underground, where they overwinter and continue developing, often germinating the following spring or even the year or two after. A complex and long seed stratification process yielded the best germination results. A well-drained growing medium was also important to sustain even temperatures and minimize bacterial and fungal damage. Other practitioners have found removing empty seeds by weight also reduces bacterial and fungal contamination caused by empty seeds decaying in stratification.

Thinning competing species to release whitebark pine

In 2017, a replicated, controlled thinning study was conducted in cutblocks in southwestern Alberta to determine the most effective competition release treatment for slow-growing whitebark pine. Blocks were harvested in the mid-1990s, scarified and regenerated mostly to lodgepole pine, but with substantial amounts of natural whitebark pine. Thinning distances, repeated in four cutblocks, were 0, 2, and 5 m radius from whitebark pine, retaining a fully stocked stand of commercial species. Data were collected on overstorey and regeneration tree species and sizes, as well as health of whitebark pines, which were all tagged. Each of the 12 plots is established as a provincial Permanent Sample Plot to facilitate remeasurement. There may be the opportunity to replicate this trial in other areas. An operational project of another 28 hectares in the same area thinned 5 m around whitebark pines.

Fire history, regeneration and health of whitebark and limber pine stands

In 2019 a project to study relationships between fire history and regeneration was conducted in the Canadian Rocky Mountains to systematically investigate anecdotal reports and regional studies suggesting whitebark and limber pine regeneration in Canada is less fire-dependent than in the U.S., possibly related to cooler, snowy conditions. A streamlined assessment of fire history evidence and categorical assessments of ground fuels and fuel types was done in nearly 250 stands planned for long-term repeat monitoring where time and resources permitted. Assessment methods may have underestimated fire evidence, especially for low severity burns because this was not a dendrochronological study. Old burns (over 20 years prior) could have been more prevalent, but recent burns (within 20 years) aligned well with available data sets.

Neither whitebark nor limber pine showed any trends in regeneration density with latitude or elevation. There was a weak relationship between regeneration density and live tree density within stands for limber pine but not for whitebark pine. Whitebark pine had higher regeneration densities than limber pine. There were no differences for health or regeneration density between burnt and unburnt sites overall. Recently burnt whitebark pine sites had higher blister rust infection rates, reflecting the increase in understorey vegetation. Unburnt sites had significantly higher whitebark pine regeneration density than sites with old burns, but only slightly less than recent burns. About half the stands of each species with regeneration had no evidence of fire, implying that fire is not essential for regeneration or recruitment in the Rocky Mountains in Canada.

Mycorrhizal effects on seedlings

In 2016, two species of Suillus mycorrhizal fungi were collected for Dr. Roland Treu of Athabasca University to inoculate whitebark pine seedlings, which improves seedling nursery growth and survival by 10 to 15%. However, no significant differences have been found in field performance following inoculation (Cripps et al. 2018). There is potential to continue this work should resources permit. Forest nurseries can operationally inoculate seedlings by applying a slurry of mycorrhizae through irrigation booms.

Charcoal/biochar effects on regeneration

In 2017, samples were collected from the upper 10 cm of soil in a burnt whitebark pine stand near Landslide Lake to support a project at University of Alberta characterizing biochemical aspects of mycorrhizal and soil characteristics in these types of stands.

DATA MANAGEMENT, INFORMATION SHARING, AND EXTENSION

Sharing and promoting knowledge spurs momentum for recovery of endangered species and ecosystems by helping the public, stakeholders, and policymakers understand their importance and the challenges that they face. Supplementary information ensures funding agencies are aware of realistic costs, limitations and timelines related to project deliverables. The slow growth and maturation of these species poses a special challenge. Most people find it hard to grasp the long consequences of impacts, and the long recovery horizon. The time scales are a poor fit for agency tracking and reporting. A seed collected now, grown into a seedling, planted and growing to maturity in its habitat, will not start to produce its own seeds for about 80 years: that is one single tree generation.

Whitebark and limber pine are well-represented in protected areas and only somewhat threatened by human impacts outside of parks, which is quite a different scenario than for most species at risk. Only active restoration measures focusing on disease resistance will ensure their long-range persistence in numbers sufficient to keep Clark's nutcracker visiting stands and planting their seeds. The multifaceted recovery actions that culminate in planting a tree take nearly a decade, and all are equally important, but most supporting activities require special expertise or training, limiting opportunities for direct public participation.

A dedicated email to reach the recovery team co-chairs, and included in current extension resources, is goa.endangeredpine@gov.ab.ca.

Data Collection

Spatial data consolidation and sharing

Locations of monitoring transects, plus trees, and restoration project areas are consolidated in the provincial Layer Manager spatial dataset, available internally to staff. Data are also submitted annually to Alberta Wildfire to identify these high value elements for fire management planning. Data is also shared with Agriculture & Forestry, Alberta Environment and Parks, Parks Canada, the U.S. Forest Service, and the Crown Managers Partnership, as well as researchers upon request. This data has also been used to support land use and resource management planning for the Castle parks management planning; implementation of the South Saskatchewan Regional Plan under the provincial Land Use Framework; forest management planning and identification of species at risk to improve forest inventory, management and operational plans; and research projects at University of Alberta, University of Northern British Columbia, and Athabasca University.

Citizen science app

In 2016 the provincial spatial analyst built a free citizen science app "Save the Pine" created in ESRI's Survey123 for recreational users and volunteers to collect location and basic access and health data on 5-needle pines. Data collected can be cached and submitted to the provincial recovery team after returning to cellular or wifi signal. However, uptake has been low and there is now better participation with individuals submitting iNaturalist records.

Plus tree and stand data collection app

Provincial staff have been using the ESRI Collector app to collect field data on mobile devices. This significantly reduced project data management time and errors, compared to non-digital options. Data were collected against background imagery linked to polygons (stands), points (trees), and other features, and backed up and managed remotely to a secure online data management hub (ArcGIS Online) after syncing to a wireless connection. Multiple crews can collect data simultaneously for the same project. Data can be exported and managed in various formats and security settings for tiered team access.

Public presentations

Invited and submitted presentations on Alberta's whitebark and limber pine recovery program have been delivered at numerous workshops, community of practice forums, academic lectures, and conferences. The theme and audience are considered carefully when developing each presentation. Hosts have included the Alberta Invasive Species Council, Alberta Native Plant Council, Alberta Forest Management Branch as well as Forest Health and Adaptation staff, Junior Forest Rangers field days, Crown Managers Partnership, seed orchard partners, provincial government spatial data community of practice, University of Alberta public lectures as well as graduate and undergraduate lectures, the general public, and naturalist groups. Specialist presentations at conferences and workshops are described in the Publications section.

Landowner, proponent, and tenure holder outreach

Alberta has engaged landowners and tenure holders by sharing information, extension materials, and results related to the trees on and near their property in conjunction with access permission requests and project referrals. All landowners and leaseholders approached to date are concerned about the status of endangered pine trees and support conservation and restoration measures, often granting access to additional areas for surveys and restoration opportunities.

Responses to project referrals overlapping or near pine habitat often spur positive engagement as proponents aim to avoid and minimize impacts, and mitigate impacts that may be unavoidable. Such projects include mining and quarry projects, powerlines, recreational facilities, expansions of existing sites, and oil and gas infrastructure. Proponents receive information to support best practices, and detailed information on plus trees and other irreplaceable installations like monitoring transects when appropriate. This has raised awareness, encouraged more proactive measures during project planning and construction, and enhanced sharing of monitoring and related data.

Locations of plus trees and habitat models have also been shared with forest tenure holders to improve outcomes for limber and whitebark pine in forest management planning and operations. Updated forest inventories incorporating this data have been used to more accurately delineate stands with unmerchantable endangered pines and remove those stands from operable areas, proactively minimizing potential impacts. Some tenure holders intend to collect location records of these species, strive to retain healthy trees in unhealthy stands, and avoid impacts to irreplaceable recovery assets such as plus trees and monitoring plots.

Indigenous community outreach

Several Alberta Indigenous Nations have reserve lands and traditional territories that overlap limber, and to a lesser extent, whitebark pine stands. After receiving permission, field crews have surveyed stands and selected plus trees there, and invited members of First Nations to join field surveys. Information is periodically shared about the recovery program, and seedlings have been provided. Work to build better connections and strengthen relationships is ongoing throughout the range of these species in Alberta.

Web extension

The recovery program is summarized in the provincial website, and includes a more detailed annual summary update available for download directly or through Open Data. https://www.alberta.ca/whitebark-and-limber-pine-recovery.aspx. Accessed January 26 2022.

A whitebark pine ESRI Story Map "Living on the Edge" was published highlighting the recovery program background and achievements. https://esrd.maps.arcgis.com/apps/Cascade/index.html?appid=d69f30908553449baef93beb7f7689e7. Accessed January 26 2022.

A poster was presented at the 2021 Crown of the Continent Forum on Fire in the Crown: https://www.crownmanagers.org/s/CMP_Poster2.pdf. Accessed January 26 2022.

An article was published on planting seedlings with Nature Conservancy of Canada volunteers: https://www.e-know.ca/regions/east-kootenay/volunteers-plant-endangered-trees-in-crowsnest-pass/?fbclid=IwAR0jiPIwbjvQiR5DuXtbmrFrPFWd9FMQVS-QuA6AUXHUKoqIhs-gsvWxbxc. Accessed January 26 2022.

An article on restoration planting was posted on the Whitebark Pine Ecosystem Foundation website and newsletter: https://whitebarkfound.org/slowly-but-surely-for-albertas-whitebark-and-limber-pine/. Accessed January 26 2022.

Alberta Environment and Parks has since archived a short 2016 blog on the recovery program.

Training

Field training is essential to collect accurate and consistent data, rigorously select plus trees, and safely collect high-quality cones and scion. At the beginning of each field season, field crews from multiple agencies cross-train together in the field to benefit from detailed hands-on field training sessions facilitated by agency experts to learn how to identify forest health issues and collect consistent, high-quality data for Parks Canada, Alberta and other partners. Training sessions specific to addressing the unique challenges of whitebark and limber pine cone collection have been developed and hosted by ArborCanada and Parks Canada, including a train-the-trainer module in 2019. These climbing and access methods have been adopted as agency Standard Operating Procedures and by the Forestry Division Occupational Health and Safety program. Field volunteers, as well as staff from other agencies and NGOs, have also been trained in the past on a case-by-case basis. In future years, training for NGOs is anticipated to support broadening capacity and empowering conservation landowners to support the program on their own conservation properties.

Publications

In 2015 updated whitebark and limber pine Alberta Species At Risk program brochures, bookmarks, stickers and magnets were produced, as well as interpretive all-weather signs in three sizes highlighting limber and whitebark pine recovery work. Signs were distributed to AEP staff to post at trailheads, along trails and at staging areas near limber and whitebark pine stands. Brochures are available electronically through the AEP Species at Risk and Government of Alberta Open Data websites, and hard copies of these materials are available upon request and provided to various offices for distribution.

Publicly accessible short articles on Alberta's recovery program have been published in Nutcracker Notes, the twice-annual publication of the Whitebark Pine Ecosystem Foundation (e.g., Krakowski 2020), which receives citations in peer-reviewed publications. Bugs & Diseases, the Alberta Forestry Division Forest Health and Adaptation newsletter published three times annually, has carried regular program and project updates. Articles describing program developments have been published in the twice-annual Tree Seed Working Group Bulletin, published by the Canadian Forest Genetics Association (Krakowski 2017, 2018, 2019).

Best practices for working with these species in Alberta are available upon request from the provincial recovery team and posted on the website of the Whitebark Pine Ecosystem Foundation of Canada. Agencies such as the High-Five Working Group of the Crown Managers Partnership, Whitebark Pine Ecosystem Foundation (Tomback et al. 2022), the Province of B.C. (Moody and Pigott 2021; Pigott et al. 2017), and Parks Canada also have developed guidelines tailored to those regions.

A summary of the provincial five-needle pine program was published in the proceedings of the 2016 international conference Forest Gene Conservation: Banking on the Future. The whitebark and limber pine habitat suitability modelling was published in the 2017 proceedings of the joint Canadian and Western Forest Genetics Associations. The Alberta limber pine program disease resistance screening for major gene resistance and multigenic resistance is published in the 2018 proceedings of the 6th International Workshop on the Genetics of Tree-Parasite Interactions: Tree Resistance to Insects and Diseases: Putting Promise into Practice (Sniezko et al. 2020), and an update is in press in the proceedings of the International Union of Forest Research Organizations (IUFRO) conference of working groups on pine stem rusts and genetics of five-needle pines (Krakowski, in press). A summary of the study on fire and regeneration was presented at the 2021 Crown Managers Partnership Forum on Fire in the Crown of the Continent (Krakowski et al. 2021). Genomics applications and tools related to major gene resistance identified in Alberta limber pine and potential applications in recovery have been published in journals (Liu et al. 2020; Sniezko et al. 2016) and presented at several conferences including the above IUFRO meeting (Sniezko et al. 2019) and the 2021 High Five II conference on high-elevation five-needle pines (Sniezko et al. 2021) where several other papers and posters relevant to the Alberta recovery program were presented. Alberta has contributed material and data for numerous other publications on the genetics and general status and recovery outlook of these species.

FUTURE DIRECTION

Because of the irregular cone crops and the labour-intensive and specialized demands of seed collection, multiple years of planning. Regular activities include: monitoring plus tree health status, collecting stand basal area and/or density data to improve characterization of stand density and delineation of critical habitat as defined in the federal recovery strategy, protecting at-risk plus trees from mountain pine beetle, identifying new plus trees, collecting scions for gene conservation in seed orchard and/or clone bank locations, keeping a continuous supply in production of seedlings to plant, and matching sites with appropriate restoration activities.

The five-year long-term monitoring program remeasurement next occurs in 2024. Given the major workload, other activities, besides fall planting, may not occur.

Based on ongoing tracking and monitoring results and new research, the recovery plan and supporting elements including seed zones and seed transfer rules may be revised from time to time to maximize program effectiveness and incorporate new knowledge. The spatial and temporal scope of the recovery program for these species means that a long term, dedicated effort among partners is essential for success.

LITERATURE CITED

Alberta Agriculture and Forestry. 2018a. Gene conservation plan for native trees of Alberta: second edition. Government of Alberta. Edmonton, AB. 112 p. https://open.alberta.ca/publications/9781460141335. Accessed 26 January 2022.

Alberta Agriculture and Forestry. 2018b. Ex situ conservation plan for forest genetic resources in Alberta. Government of Alberta. 62 p. https://open.alberta.ca/publications/9781460138793. Accessed 26 January 2022.

Alberta Whitebark and Limber Pine Recovery Team. 2014a. Alberta whitebark pine recovery plan 2013-2018. Alberta Environment and Sustainable Resource Development. Alberta Species at Risk Recovery Plan no. 34, Edmonton, AB. 71 p. https://open.alberta.ca/publications/9781460111000#-summary. Accessed 26 January 2022.

Alberta Whitebark and Limber Pine Recovery Team. 2014b. Alberta limber pine recovery plan 2014-2019. Alberta Environment and Sustainable Resource Development. Alberta Species at Risk Recovery Plan no. 35, Edmonton, AB. 61 p. https://open.alberta.ca/publications/978146011848#summary. Accessed 26 January 2022.

Barringer, LE, DF Tomback, MB Wunder, and ST McKinney. 2012. Whitebark pine stand condition, tree abundance, and cone production as predictors of visitation by Clark's Nutcracker. PLOS ONE 7(5):e37663. doi: 10.1371/journal.pone.003766.

Cardinal, E, B Shepherd, J Krakowski, C Schwarz, and J Stirret-Wood. 2021. Verbenone and green-leaf volatiles reduce whitebark pine mortality in a northern range-expanding mountain pine beetle outbreak. Canadian Journal of Forest Research. https://doi.org/10.1139/cjfr-2021-0120.

Clason, AJ. 2013. A tactical plan for whitebark pine restoration in the Omineca region. Report for B.C. Ministry of Forests, Lands and Natural Resource Operations. Smithers, B.C. 77 p. http://sernbc.ca/uploads/11/WBP_management_in_Omineca_Final_draft.pdf. Accessed 26 January 2022.

COSEWIC [Committee on the Status of Endangered Wildlife in Canada]. 2010. COSEWIC assessment and status report on the whitebark pine *Pinus albicaulis* in Canada. Committee on the Status of Endangered Wildlife in Canada. Ottawa, ON. 44 p. https://publications.gc.ca/collections/collection_2011/ec/CW69-14-612-2010-eng.pdf. Accessed 26 January 2022.

COSEWIC [Committee on the Status of Endangered Wildlife in Canada]. 2014. COSEWIC assessment and status report on the limber pine *Pinus flexilis* in Canada. Committee on the Status of Endangered Wildlife in Canada. Ottawa, ON. 49 p. https://wildlife-species.canada.ca/species-risk-registry/virtual_sara/files/cosewic/sr_Limber%20Pine_2014_e.pdf. Accessed 26 January 2022.

Cripps, CL, G Alger, and R Sissons. 2018. Designer niches promote seedling survival in forest restoration: a 7-year study of whitebark pine (*Pinus albicaulis*) seedlings in Waterton Lakes National Park. Forests 9(8):477. DOI: 10.3390/f9080477.

Environment Canada. 2017. Recovery strategy for whitebark pine (*Pinus albicaulis*) in Canada [Proposed]. Species at Risk Act Recovery Strategy Series. Environment Canada, Ottawa, ON. 54 p. https://www.registrelep-sararegistry.gc.ca/virtual_sara/files/plans/rs_whitebark_pine_e_proposed.pdf. Accessed 26 January 2022.

Jenkins, MB, AW Schoettle, JW Wright, KA Anderson, J Fortier, L Hoang, T Incashola Jr., RE Keane, J Krakowski, D LaFleur, S Mellmann-Brown, ED Meyer, S Pete, K Renwick, and RA Sissons. [in press]. Restoring a keystone forest tree: a plan for the restoration of whitebark pine (*Pinus albicaulis* Engelm.) in the Crown of the Continent ecosystem. Forest Ecology and Management. Special issue: ecology and restoration of high-elevation five-needle white pines. Edited by DF Tomback, RE Keane, and RA Sniezko.

Keane, RE, L Holsinger, MF Mahalovich, and DF Tomback. 2017. Restoring whitebark pine in the face of climate change. General Technical Report RMRS-GTR-361. Fort Collins, CO: U.S. Department of Agriculture, Forest Service, Rocky Mountain Research Station. 133 p.

Krakowski, J. 2017. Forest gene conservation programs in Alberta, Canada. In Gene conservation of tree species—banking on the future, Gen. Tech. Rep. PNW-GTR-963, technical coordation by RA Sniezko, G Man, V Hipkins, K Woeste, D Gwaze, JT Kliejunas, and BA McTeague, 170-173. Portland, OR: U.S. Department of Agriculture, Forest Service, Pacific Northwest Research Station.

Krakowski, J. 2017. One piece of the puzzle: 5-needle pine ex situ conservation in Alberta. Canadian Forest Genetics Association, Tree Seed Working Group News Bulletin 65:11-14. https://www2.gov.bc.ca/assets/gov/farming-natural-resources-and-industry/forestry/tree-seed/tree-seed-centre/twsgnewsbulletin65a.pdf. Accessed 26 January 2022.

Krakowski, J. 2018. Alberta: interim genetic resource management rules approved for whitebark pine and limber pine. Canadian Forest Genetics Association, Tree Seed Working Group News Bulletin 66:11-12. https://www2.gov.bc.ca/assets/gov/farming-natural-resources-and-industry/forestry/tree-seed/tree-seed-centre/tswgnewsbulletin66.pdf. Accessed 26 January 2022.

Krakowski, J. 2019. Alberta's 2018 Endangered pine seed collection and restoration planting for recovery. Canadian Forest Genetics Association, Tree Seed Working Group News Bulletin 68:20-22. https://www2.gov.bc.ca/assets/gov/farming-natural-resources-and-industry/forestry/tree-seed/tree-seed-centre/tswg-bulletins/tswg_news_bulletin_68.pdf. Accessed 26 January 2022.

Krakowski, J. 2020. Update on Alberta monitoring. Nutcracker Notes 38:25-26. https://drive.google.com/file/d/1DASNeqxKg1IPbchyCf_0FfuPflVCPEUl/view?usp=sharing. Accessed 26 January 2022.

Krakowski, J, J Gould, M Berkhout, R Gutsell, and P Melnick. 2021. Does fire really matter? In: Whitebark and limber pine regeneration and fire in the Canadian Rockies. Crown Managers Partnership Forum 2021: Fire in the Crown of the Continent. https://www.crownmanagers.org/s/CMP_Poster2.pdf. Accessed 26 January 2022.

Krakowski, J, R Kite, and A Blythe. 2017. In the right place: habitat suitability models for endangered whitebark pine and limber pine to support recovery and management. In: Forest genetics 2017: health and productivity under changing environments, 61. Edmonton, AB: A Joint Meeting of WFGA and CFGA. http://fsl.orst.edu/wfga/proceedings/2017_Proceedings.pdf. Accessed 26 January 2022.

Krakowski, J, R Gutsell, B Jones. [in press] Reach for the top: implementing whitebark and limber pine recovery in Alberta. In IUFRO 2019 joint conference: genetics of five-needle pines & rusts of forest trees. Compiled by J Krakowski, W Strong, and RA Sniezko. B.C. Ministry of Forests, Lands and Natural Resource Operations and Rural Development, Tech. Rep. Series.

Krakowski, J, R Heinzelmann, T Ramsfield, A Benowicz, R Hamelin, and C Myrholm. 2021. First report of *Dothistroma septosporum* on *Pinus flexilis* in Canada. Canadian Plant Disease Survey 101:176-178.

Liu, J-J, RA Sniezko, R Sissons, J Krakowski, G Alger, AW Schoettle, H Williams, A Zamany, RA Zitomer, and A Kegley. 2020. Association mapping and development of marker-assisted selection tools for the resistance to white pine blister rust in the Alberta limber pine populations. Frontiers in Plant Science 11:557672. DOI: 10.3389/fpls.2020.557672.

Mahalovich, MF, and GA Dickerson. 2004. Whitebark pine genetic restoration program for the Intermountain West (United States). In Proceedings on breeding and genetic resources of five-needle pines: growth, adaptability and pest resistance. Edited by RA Sniezko, S Samman, SE Schlarbaum, and HB Kriebel, 181-187. Fort Collins, CO: U.S. Department of Agriculture, Forest Service RMRS-P-32, Rocky Mountain Research Station.

McDermid, GJ, and IU Smith. 2008. Mapping the distribution of whitebark pine (*Pinus albicaulis*) in Waterton Lakes National Park using logistic regression and classification tree analysis. Canadian Journal of Remote Sensing 34:356-366.

McKinney, ST CE Fiedler, and DF Tomback. 2009. Invasive pathogen threatens bird-pine mutualism: implications for sustaining a high-elevation ecosystem. Ecological Applications 19:597-607.

Moody, R, and D Pigott. 2021. Best management practice for whitebark pine (*Pinus albicaulis*). Report to Species At Risk Recovery Branch, B.C. Ministry of Forests, Lands and Natural Resource Operations and Rural Development. 109 p. https://www2.gov.bc.ca/assets/gov/environment/natural-resource-stewardship/best-management-practices/whitebark_pine_bmp.pdf. Accessed 26 January 2022.

Parks Canada Agency. 2017a. Multi-species action plan for Banff National Park of Canada. Species At Risk Act Action Plan Series. Parks Canada Agency, Ottawa, ON. 27 p. https://www.canada.ca/en/environment-climate-change/services/species-risk-public-registry/action-plans/banff-national-park-multi-species.html. Accessed 26 January 2022.

Parks Canada Agency. 2017b. Multi-species action plan for Waterton Lakes National Park of Canada and Bar U Ranch National Historic Site of Canada. Species At Risk Act Action Plan Series. Parks Canada Agency, Ottawa, ON. 24 p. Accessed 26 January 2022. https://www.canada.ca/en/environment-climate-change/services/species-risk-public-registry/action-plans/waterton-lakes-national-park-historic-site.html. Accessed 26 January 2022.

Parks Canada Agency. 2017c. Multi-species action plan for Jasper National Park of Canada. Species At Risk Act Action Plan Series. Parks Canada Agency, Ottawa, ON. 30 p. https://publications.gc.ca/collections/collection_2017/pc/CW69-21-41-2017-eng.pdf. Accessed 26 January 2022.

Perkins, DL, RE Means, and AC Cochrane. 2016. Conservation and management of whitebark pine ecosystems on Bureau of Land Management lands in the western United States. Denver, CO: Tech. Ref. 6711-1. Bureau of Land Management. 93 p. https://www.blm.gov/sites/blm.gov/files/documents/files/Library_BLMTechnicalReference6711-01.pdf. Accessed 26 January 2022.

Pigott, D, R Moody, and A Clason. 2015. Promoting whitebark pine recovery in British Columbia. Report to B.C. Ministry of Forests, Lands and Natural Resource Operations. 108 p. https://sernbc.ca/uploads/14/Promoting_Whitebark_Pine_Recovery_in_BC.pdf. Accessed 26 January 2022.

Robb, L. 2020. Seed matters 2 - extracting and handling five-needle pine seed in Alberta. Alberta Agriculture and Forestry, Government of Alberta. 13 p. https://open.alberta.ca/publications/9781460146309. Accessed 26 January 2022.

Schoettle, AW, KS Burns, J Krakowski, ST McKinney, KM Waring, and DF Tomback. [in press]. Integrating forest health condition and species adaptive capacity to infer and affect future outcomes of the high-elevation five-needle pines. Forest Ecology and Management. Special issue: ecology and restoration of high-elevation five-needle white pines. Edited by DF Tomback, RE Keane, and RA Sniezko.

Shepherd, B, B Jones, R Sissons, J Cochrane, J Park, CM Smith, and N Stafl. 2018. Ten years of monitoring illustrates a cascade of effects of white pine blister rust and focuses whitebark pine restoration in the Canadian Rocky and Columbia Mountains. Forests 9(3):138. https://doi.org/10.3390/f9030138.

Smith, CM, DW Langor, C Myrholm, J Weber, C Gillies, and J Stuart-Smith. 2013a. Changes in white pine blister rust infection and mortality in limber pine over time. Canadian Journal of Forest Research 43:919-928.

Smith, CM, B Shepherd, C Gillies, and J Stuart-Smith. 2013b. Changes in blister rust infection and mortality in whitebark pine over time. Canadian Journal of Forest Research 43(1):90-96.

Smith, CM, B Wilson, S Rasheed, RC Walker, T Carolin, and B Shepherd. 2008. Whitebark pine and white pine blister rust in the Rocky Mountains of Canada and northern Montana. Canadian Journal of Forest Research 38(5):982-995.

Sniezko, RA, A Kegley, J Krakowski, G Alger, and J-J Liu [in press]. White pine blister rust resistance programs for Alberta limber and whitebark pine recovery. In H5II: the second conference on the research and management of high-elevation five needle pines. Edited by CM Smith, P Achuff, ST McKinney, and M.P. Murray. Humboldt State University Press.

Sniezko, RA, R Danchok, DP Savin, J-J Liu, and A Kegley. 2016. Genetic resistance to white pine blister rust in limber pine (*Pinus flexilis*): major gene resistance in a northern population. Canadian Journal of Forest Research 46:1173-1178.

Sniezko, RA, M Lewien, J Krakowski, R Sissons, G Alger, J-J Liu, R Zitomer, and A Kegley. [in press]. Finding and using genetic resistance to white pine blister rust in Alberta populations of limber pine (*Pinus flexilis*). In IUFRO 2019 joint conference: genetics of five-needle pines & rusts of forest trees. Compiled by J Krakowski, W Strong, and RA Sniezko. B.C. Ministry of Forests, Lands and Natural Resource Operations and Rural Development, Tech. Rep. Series.

Sniezko, RA, J Krakowski, R Sissons, R Danchok, A Kegley, and DP Savin. 2020. Genetic resistance to blister rust in declining limber pine (*Pinus flexilis*) in Alberta-the path to restoration? In Proceedings of the 6th international workshop on the genetics of host-parasite interactions: tree resistance to insects and diseases: putting promise into practice,

General Technical Report SRS-252, edited by CD Nelson, JL Koch, and RA Sniezko, 158. Asheville, NC: U.S. Department of Agriculture, Forest Service, Southern Research Station.

Tomback, DF, RE Keane, WW McCaughey, and CM Smith. 2005. Methods for surveying and monitoring whitebark pine for blister rust infection and damage. Whitebark Pine Ecosystem Foundation, Missoula, MT. 35 p. http://whitebarkfound.org/wp-content/uploads/2013/10/Methods-for-Surveying-and-Monitoring-Whitebark-Pine-for-Blister-Rustx.pdf. Accessed 26 January 2022.

Tomback, DF, RE Keane, AW Schoettle, RA Sniezko, MB Jenkins, DR Nelson, AD Bower, CR DeMastus, E Guiberson, J Krakowski, MP Murray, ER Pansing, and J Shamhart. [in press]. Tamm review: current and recommended management practices for the restoration of whitebark pine (*Pinus albicaulis* Engelm.), a threatened high-elevation Western North American forest tree. Forest Ecology and Management. Special issue: ecology and restoration of high-elevation five-needle white pines, edited by DF Tomback, RE Keane, and RA Sniezko.

Wilson, BC, and JG Stuart-Smith. 2001. Whitebark pine conservation for the Canadian Rocky Mountain National Parks. Cordilleran Ecological Research Report KNP01-01, Winlaw, B.C. 30 p.

Alpine Treeline Ecotones are Potential Refugia for a Montane Pine Species Threatened by Bark Beetle Outbreaks

Colin T. Maher, <cmbristlecone@gmail.com>,WA Franke College of Forestry and Conservation, The University of Montana; **Constance I. Millar**, <connie.millar@usda.gov>, Pacific Southwest Research Station, USDA Forest Service; **David L.R. Affleck**, <david.affleck@umontana.edu>, WA Franke College of Forestry and Conservation, The University of Montana; **Robert E. Keane**, <robert.e.keane@usda.gov>, Rocky Mountain Research Station, Missoula Fire Science Laboratory, USDA Forest Service; **Anna Sala**, <asala@mso.umt.edu>, Division of Biological Sciences, The University of Montana; **Claudine Tobalske**, <claudine.tobalske@mso.umt.edu>, Spatial Analysis Lab, Montana Natural Heritage Program, The University of Montana; **Andrew J. Larson**, <a.larson@umontana.edu>, WA Franke College of Forestry and Conservation, The University of Montana; **Cara R. Nelson**, <cara.nelson@umontana.edu>, WA Franke College of Forestry and Conservation, The University of Montana.

ABSTRACT

Warming-induced mountain pine beetle (*Dendroctonus ponderosae*; MPB) outbreaks have caused extensive mortality of whitebark pine (*Pinus albicaulis*; WBP) throughout the species' range. In high mountains, WBP cross alpine treeline ecotones (ATEs) where growth forms transition from trees to shrub-like krummholz—some of which survived recent outbreaks. This observation motivated the hypothesis that ATEs are refugia for WBP because krummholz escape MPB attack and can reproduce. To test this hypothesis, we surveyed WBP mortality along transects from the ATE edge (locally highest krummholz WBP) downslope into the forest and—to distinguish if survival mechanisms are unique to ATEs—across other forest ecotones (OFEs) from the edge of WBP occurrence into the forest. We replicated this design at 10 random sites in the US Northern Rocky Mountains. We also surveyed reproduction at 3 ATE sites. Mortality was nearly absent in upper ATEs (mean ± 1 s.e. % dead across all sites of 0.03 ± 0.03% 0-100 m from the edge and 14.1 ± 1.7% 100-500 m from the edge) but was above 20% along OFEs (21.4 ± 5.2% 0-100 m and 32.4 ± 2.7% 100-500 m). We observed lower reproduction in upper ATEs (16 ± 9.9 cones·ha-1 and 12.9 ± 5.3 viable seeds·cone-1 0-100 m) versus forests below (317.1 ± 64.4 cones·ha-1 and 32.5 ± 2.5 viable seeds·cone-1 100-500 m). Uniquely high survival of krummholz and small trees in the ATE indicate they escaped MPB attack, supporting the refugia hypothesis. However, low reproduction suggests ATE refugia function over longer time periods.

Keywords: boundary, climate change refugia, edge, mountain pine beetle, *Pinus albicaulis*, tree mortality, whitebark pine

INTRODUCTION

The upper elevation boundaries of whitebark pine (*Pinus albicaulis* Emgelm.; hereafter WBP) forests—alpine treeline ecotones (ATEs)—typically contain shrub-like krummholz growth forms (Arno 1984). These ATEs are characterized by gradients from forests with tall trees to areas with dispersed short-stature trees and finally to the tree species limit at the upper edge of the ecotone—the krummholz zone (Griggs 1946; Körner and Paulsen 2004). In WBP populations impacted by mountain pine beetle (MPB) outbreaks, ATE habitats are hypothesized to serve as refugia: Logan et al. (2010) and Macfarlane et al. (2013) observed that live WBP krummholz in ATEs persisted above forests with extensive MPB-caused mortality (figure 1). Macfarlane et al. (2013) hypothesized that "long-term survival of the species likely resides in the [krummholz] growth form found throughout the ecosystem near treeline, because it is too small for beetles to attack..."

Refugia for WBP from MPB outbreaks is plausible in ATEs for two principal reasons. First, krummholz WBP are not genetically distinct from tree WBP (Rogers et al. 1999). Thus, krummholz or their offspring would likely grow as trees in a milder environment. Second, MPBs are known to prefer trees with diameters greater than 10-15 cm (Cole and Amman 1969). The small stems and contorted shapes of krummholz and other treeline growth forms may underpin mechanisms that could maintain WBP refugia from MPB in ATE habitats.

Despite the plausibility of the ATE refugia hypothesis, there are alternate explanations for a pattern of low WBP mortality above treeline. For instance, ATEs may share with other forest ecotones (OFEs) key attributes that affect MPB spread and impacts. Changes in structure near forest boundaries are known to modulate effects of herbivores (i.e., "edge effects"), depending on the herbivore's behavior (Cadenasso and Pickett 2000). Interruption of MPB pheromone signals by wind may occur at all forest boundaries—pheromone plumes are diluted by circulation (Thistle et al. 2004). Because krummholz is unique to ATEs, an examination of MPB impacts across OFEs should provide a first approximation of whether the mechanisms of survival are related to growth form.

In this research we identify possible refugia from climate change effects for a montane tree species with a focus on understanding the disturbance-related mechanisms that maintain the refugia. The specific goals of this research were to verify that WBP populations in the ATE did survive recent MPB outbreaks, to distinguish between plausible mechanisms of survival in the ATE, and to examine overall survival in post-MPB outbreak WBP forests in the U.S. Northern Rocky Mountains. We asked:

(1) Are mortality rates of WBP in upper ATEs lower than in subalpine forest interiors and how does this mortality gradient differ from OFEs?

(2) What is the overall post-outbreak survival status of WBP populations in the US Northern Rocky Mountains?

METHODS

We used GIS layers of MPB-caused forest mortality, alpine vegetation type, and WBP occurrence to create a sampling frame of possible field site locations using ArcMap (ESRI 2010). We used a detailed 2008 aerial survey focused specifically on MPB-caused mortality of WBP in the Greater Yellowstone Ecosystem (Macfarlane et al., 2013) and Maps of MPB-caused mortality in the US states of Idaho, Montana, and Wyoming from the USDA Forest Service's Forest Health Protection aerial Insect and Disease Surveys (IDS; Forest Health Protection, 2014). We filtered mortality data to include only polygons where MPB-caused mortality was observed in WBP and where these areas overlapped with or were adjacent to alpine vegetation in USGS GAP land cover (US Geological Survey Gap Analysis Program (GAP) 2011). The final sampling frame was a ~7,480 km^2 area.

We selected field sampling locations within the sampling frame by randomly placing 10 points in ArcGIS (n = 10). Near each sampling point, we initiated two transects: one at the nearest ATE edge and one at the nearest OFE edge. We defined ecotone "edges" as the last WBP bordering the alpine or other non-forest vegetation). ATE edges were the highest elevation 'outpost' krummholz visible from satellite imagery. We defined OFEs as WBP forest boundaries with non-forest openings formed by avalanche paths, forest-meadow interfaces, geologic and topographic features, or other forest margins not associated with elevation-related boundaries. ATE transects were oriented downslope following the elevational transition from krummholz to forest. OFE transects were oriented perpendicular to the WBP edge extending into the forest, regardless of slope direction. All transects were 50 m wide and 500 m long. Transects were divided into twenty contiguous 50 m by 25 m rectangular plots (the unit of analysis). Transect plot outlines were ported to a GPS device to guide field sampling.

Within each of the ATE and OFE transect plots (n = 20 plots each) at our 10 sites, we tallied WBP by status (live or dead) and by growth form (upright trees or krummholz plants). Trees were defined as WBP stems that were at least 3 m tall (regardless of stem diameter). Fused stems were considered separate if their junction was below 1.4 m from the ground. Krummholz were defined as WBP with crowns that were ≥ 1 m across and at least as wide as they were tall but were < 3 m tall. We used these classifications to calculate the proportion of tree-form WBP in each transect plot (n WBP trees/ (n WBP trees + WBP krummholz)). We recorded cause of death for all dead WBP that retained bark—we assumed

Figure 1. Survival and cone production in whitebark pine near treelines. a) Surviving whitebark pines near treeline (green band at upper edge of forest) with extensive mortality in the subalpine forest below (grey trees). Tobacco Root Mountains, Montana, U.S.A. b) Abundant 2018 cone crop on small-diameter and krummholz whitebark in the Pioneer Mountains, Montana, USA. c) 2020 current year and previous year cones on krummholz whitebark in the Sierra Nevada Mountains, California, USA. Photo credits: C. T. Maher. (Maher et al. 2021)

that WBP without bark died long before the most recent MPB outbreaks. MPB was determined as the cause of death by peeling away bark and identifying one or more of the following: j-shaped galleries, pitchout evidence, or frass (according to USDA Forest Service Common Stand Exam criteria; USDA Forest Service, 2016). Field sampling was completed in July-October 2015 and July-October 2016. We estimated pre-outbreak density of WBP as the sum of both live and dead individuals (krummholz and trees) in each transect plot.

Analyses and further methods are described in Maher et al. (2021).

RESULTS

MPB-killed WBP were almost non-existent in upper ATEs, with mortality becoming more common with increased distance from the edge (i.e., downslope into the subalpine forest; figure 2). Specifically, we observed only one krummholz at one site (overall mean ± 1 s.e. of 0.03 ± 0.03 % of stems across all sites) that had been killed by MPB within 100 m from ATE edges, and 14.1 ± 1.7 % dead between 100-500 m. In contrast with ATEs, mortality in OFEs extended to the ecotone edge and was higher overall than in ATEs: we observed 21.4 ± 5.2 % MPB-killed within 100 m from OFE edges and 32.4 ± 2.7 % between 100-500 m, although some sites had higher mortality near OFE edges (75-100 %). These patterns were reflected by a significant interactive effect on total MPB-killed WBP between distance from edge and ecotone type (χ^2 = 45.5, df = 2, P < 0.0001).

We found an overall higher density of living (mean ± 1 s.e.; 286.3 ± 72.0 WBP·ha^{-1}) versus MPB-killed (62.4 ± 16.4 WBP·ha^{-1}) WBP on a whole transect basis across our study region when including both ecotone types (two-sample Wilcoxon signed rank test, P = 0.002; figure 2). This overall effect was driven mainly by differences in live vs MPB-killed density in ATEs (273.2 ± 73.2 vs 40.4 ± 8.4 WBP·ha^{-1}; P = 0.002); there was no significant difference in OFEs (299.4 ± 128.3 vs 84.4 ± 31.0 WBP·ha^{-1}; P = 0.2).

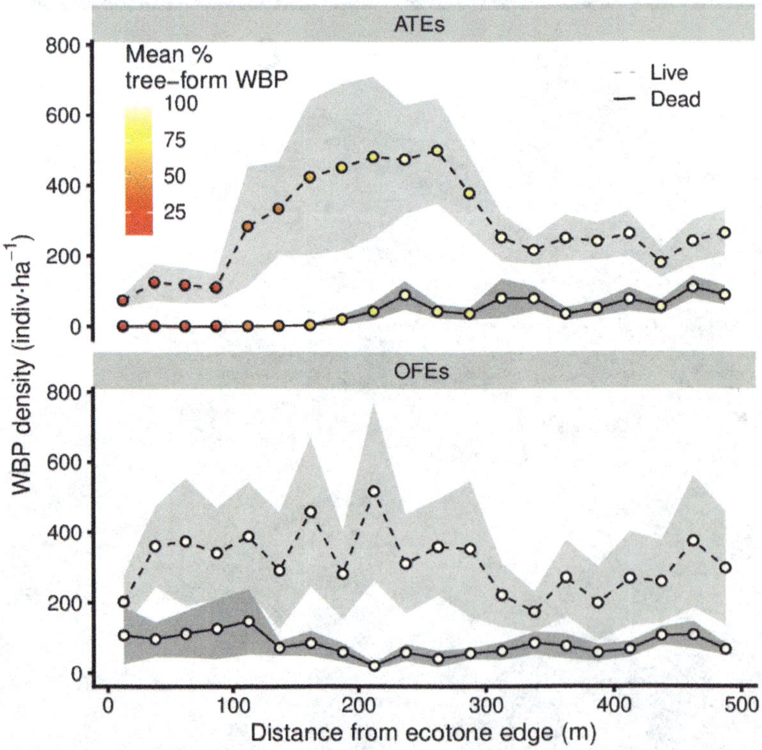

Figure 2. Mean density of living (dashed black lines) and mountain pine beetle-killed (solid black lines) whitebark pine (WBP) along alpine treeline ecotones (ATEs; top plot) and other forest ecotones (OFEs; bottom plot) at n = 10 sites. This data excludes seedlings and saplings. Symbol color represents the mean percentage of tree-form (> 3 m tall) whitebark pine (in contrast with krummholz growth forms; at least 1 m wide and 1 m tall and as wide or wider than tall) at each transect position: white represents 100 % tree-form (0 % krummholz) and red 0 % tree-form (100% krummholz). Grey bands represent ± 1 standard error of the mean. (Maher et al. 2021)

CONCLUSIONS & MANAGEMENT IMPLICATIONS FOR WHITEBARK PINE

ATE habitats afford potential climate change refugia for WBP from MPB attacks. The krummholz and stunted trees that exist in these high mountain populations are a mechanism by which WBP could remain in a landscape over long time scales as climate change causes fluctuations in disturbance regimes. Furthermore, MPB-impacted forests with surviving WBP may retain populations, due in part to the growth potential of small-diameter individuals. These populations could persist because insect outbreaks tend to be episodic—i.e., the pressure on tree populations might not be constant into the future, allowing some recovery between outbreaks. Further, there may not be a need for management intervention in some locations, and some management actions may be harmful, e.g., the use of prescibed fire where survivors are smaller trees, saplings, and seedlings. While uncertainty remains about the future of survivors in subalpine forests below the treeline, WBP in ATEs may allow for population persistence and may eventually contribute to population recovery in other habitats.

Our findings suggest that ATE habitats should be considered valuable attributes of management units containing WBP. While many common management actions (e.g., silviculture or fuels treatments) might not be applicable directly in ATE habitats, planting WPBR-resistant seedlings or directly sowing seeds is possible (Keane 2018), if appropriate given the sensitivity of these environments or their status as federally designated Wilderness (Tomback 2014). Given that planting WPBR-resistant seedlings is projected to have benefits over centuries, not decades (Keane et al. 2017), it makes sense to ensure that resistant genotypes are represented in treeline environments where WBP have the best chance of surviving future MPB outbreaks. Additionally, because treelines are an important front of climate change, promoting and maintaining already MPB-resistant ATE populations of WBP that are also resistant to WPBR will be an important aspect of ensuring the species' success into the future.

ACKNOWLEDGEMENTS

Funding for this research was provided by grants from the Montana Institute on Ecosystems, the Whitebark Pine Ecosystem Foundation, and the Associated Students of the University of Montana as well as the NIFA McIntire-Stennis Cooperative Forestry Program [project accession no. 225109] from the USDA National Institute of Food and Agriculture. We thank David Foushee and Kevin Egan at the USDA Forest Service Coeur d'Alene Nursery for seed x-rays and for providing guidance on identifying viable seed from x-ray imagery. Wally Macfarlane generously shared GIS data on mountain pine beetle impacts in WBP forests of the Greater Yellowstone Ecosystem. We thank Cullen Weisbrod, Jesse Bunker, Alyssa Hands, Haley Wiggins, Michael Fazekas, and Danica Born-Ropp for their help field sampling.

REFERENCES

Arno, SF. 1984. Timberline: Mountain and Arctic Forest Frontiers. Seattle, WA, USA: Mountaineers Books.

Cadenasso ML, and STA Pickett . 2000. Linking forest edge structure to edge function: mediation of herbivore damage. J Ecol. 88(1):31–44. doi.org/10.1046/j.1365-2745.2000.00423.x.

Cole, W. E., and GD Amman. 1969. Mountain pine beetle infestations in relation to lodgepole pine diameters. Research Note INT-95. USDA Forest Service Intermountain Range and Experiment Station, Ogden, Utah, USA.

ESRI (Environmental Systems Resource Institute). 2010. ArcGIS. ESRI, Redlands, California, USA.

Forest Health Protection. 2014. Insect and disease detection survey database (IDS). Digital Data. Fort Collins, CO: U.S. Department of Agriculture, Forest Service, Forest Health Technology Enterprise Team. http://foresthealth.fs.usda.gov/ids.

Griggs, RF. 1946. The timberlines of Northern America and their Interpretation. Ecology. 27(4):275–289. doi.org/1933539.

Keane, RE. 2018. Managing wildfire for whitebark pine ecosystem restoration in western North America. Forests. 9(10):648. doi.org/10.3390/f9100648.

Keane, RE, LM Holsinger, MF Mahalovich, and D F Tomback. 2017. Restoring whitebark pine ecosystems in the face of climate change. General Technical Report RMRS-GTR-36. USDA Forest Service, Rocky Mountain Research Station, Fort Collins, CO, USA. doi.org/10.2737/RMRS-GTR-361.

Körner C, and J Paulsen. 2004. A world-wide study of high altitude treeline temperatures. J Biogeogr. 31(5):713–732. doi.org/10.1111/j.1365-2699.2003.01043.x.

Logan JA, WW MacFarlane, and L Willcox. 2010. Whitebark pine vulnerability to climate-driven mountain pine beetle disturbance in the Greater Yellowstone Ecosystem. Ecol Appl. 20(4):895–902. doi.org/10.1890/09-0655.1.

Macfarlane WW, JA Logan, and WR Kern. 2013. An innovative aerial assessment of Greater Yellowstone Ecosystem mountain pine beetle-caused whitebark pine mortality. Ecol Appl. 23(2):421–437. doi.org/10.1890/11-1982.1.

Maher CT, CI Millar, DLR Affleck, RE Keane, A Sala, C Tobalske, AJ Larson, and CR Nelson. 2021. Alpine treeline ecotones are potential refugia for a montane pine species threatened by bark beetle outbreaks. Ecol Appl. 31(3):e02274. doi.org/10.1002/eap.2274.

Rogers, DL, CI Millar, and RD Westfall. 1999. Fine-scale genetic structure of whitebark pine (Pinus albicaulis): associations with watershed and growth form. Evolution 53:74–90. doi.org/10.1111/j.1558-5646.1999.tb05334.x.

Thistle, HW, H Peterson H, G Allwine B Lamb, T Strand, EH Holsten, and PJ Shea. 2004. Surrogate pheromone plumes in three forest trunk spaces: Composite statistics and case studies. For Sci. 50(5):610–625. doi.org/10.1093/forestscience/50.5.610.

Tomback DF. 2014. The Wilderness Act and whitebark pine. Nutcracker Notes. 27(Winter/Spring 2014-2015):1–2. whitebarkfound.org/wp-content/uploads/2015/01/Nutcracker-Notes-27.pdf.

US Geological Survey Gap Analysis Program (GAP). 2011. National land cover, Version 2. www.usgs.gov/core-science-systems/science-analytics-and-synthesis/gap/science/land-cover-data-download?qt-science_center_objects=0#qt-science_center_objects.

USDA Forest Service. 2016. Region one vegetation classification, mapping, inventory and analysis report: common stand exam and inventory and monitoring field guide. Rep 16-5 v60. www.fs.fed.us/nrm/fsveg/index.shtml.

Winter Damage is More Important than Summer Temperature for Maintaining the Krummholz Growth Form Above Alpine Treeline

Colin T. Maher, <cmbristlecone@gmail.com>,WA Franke College of Forestry and Conservation, The University of Montana; **Cara R. Nelson**, <cara.nelson@umontana.edu> WA Franke College of Forestry and Conservation, The University of Montana; **Andrew J. Larson**, <a.larson@umontana.edu> WA Franke College of Forestry and Conservation, The University of Montana.

ABSTRACT

Cool summer air temperatures are currently thought to be the limiter of upright tree growth at alpine treelines globally. However, winter damage has long been recognized as a shaping force near alpine treelines. To distinguish between effects of growing season temperature, winter damage, and their interaction on preventing upright growth, we conducted an experiment on low-growing krummholz growth forms of *Pinus albicaulis* over summer and winter 2015-2016 at 10 sites in the Tobacco Root Mountains, Montana, USA. We experimentally manipulated four factors using a fully crossed design: shoot position (natural low position in the krummholz mat vs. propped up above the krummholz mat), summer warming (warmed vs. ambient), winter exposure (sheltered vs. exposed), and elevation position (local high vs. low krummholz limits). We also conducted an observational study of the climatic conditions associated with natural stem establishment. Winter-exposed propped shoots experienced higher mortality (10-50%) than sheltered propped shoots and shoots within the krummholz mat, whether caged or not (0-10%). Summer warming had little influence on shoot mortality. Surviving mat shoots had marginally higher growth rates than surviving propped shoots during the early growing season 2016. Natural stem establishment was associated with warmer summer temperatures, but also warmer winter temperatures, lower winter wind speeds, and lower snowpack. Our results indicate that winter damage plays a more important role than growing season temperature in maintaining krummholz. Warming may increase opportunities for shoot establishment but continuing upright growth in the krummholz zone may also require reductions in damaging winter wind and snow.

Keywords: growth limitation, krummholz flags, *Pinus albicaulis*, plant-climate interactions, snow transport, treeline ecotone, whitebark pine, wind

INTRODUCTION

Alpine treeline ecotones, where upright trees transition to low-growing alpine vegetation, are common in mountain ecosystems around the world. Similar growing season temperatures at alpine treelines globally suggest that temperature limitation of growth in upright stems could be the ultimate reason trees fail to grow at high-elevations (Körner 1998, Hoch and Körner 2003). At some treeline ecotones, however, tree species grow as shrub-like krummholz, which extends the species range above the treeline (the limit of upright stems; Körner and Paulsen 2004). Although low-growing krummholz may experience warmer growing seasons than upright stems, winter exposure to desiccation and damage from wind-transported snow is also known to harm stems that emerge from the snowpack (Hadley and Smith 1983).

Our objective was to identify the processes that maintain the krummholz growth form. We addressed this objective with two complimentary observational and experimental approaches. Specifically, we asked: 1) What climatic conditions were associated with the establishment of naturally occurring emergent stems from krummholz? 2) How do growing season temperature and winter damage affect survival of shoots within and above krummholz, and do effects vary with elevation?

METHODS

This study was conducted in the Tobacco Root Mountains located in southwestern Montana, USA. We used *Pinus albicaulis* Engelm. krummholz as a study system. P. albicaulis commonly forms krummholz above alpine treelines throughout its range. Genetic analysis indicates that growth form differences in the species represent phenotypic plasticity, not local differentiation (Rogers et al. 1999), likely resulting from dispersal across treeline ecotones by Clark's nutcracker

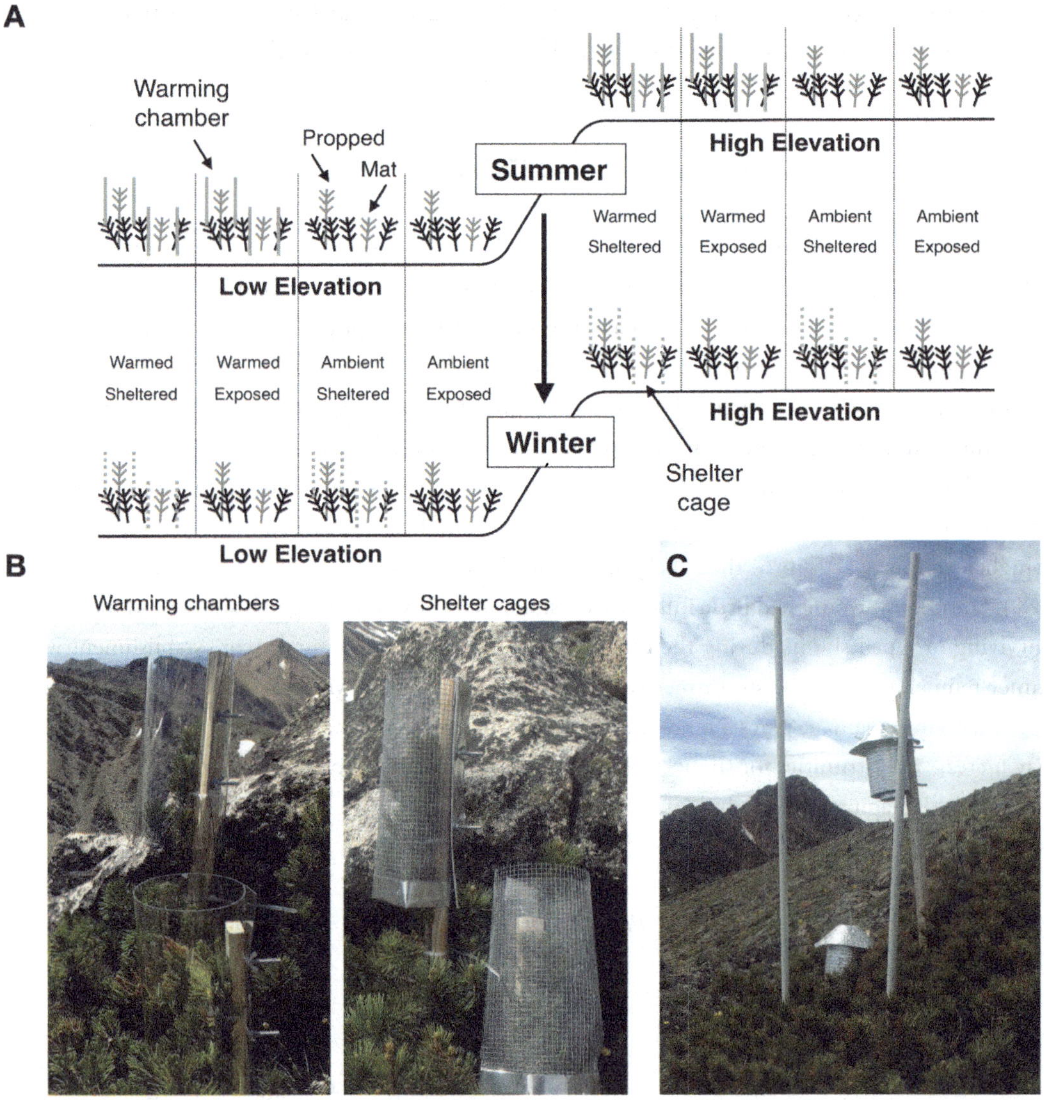

Figure 1. A) Design of krummholz shoot experiment showing four factors: shoot position (propped vs. mat), late summer warming (warmed vs. ambient), winter shelter (sheltered vs. exposed), and elevation position (local lowest krummholz vs. highest krummholz). Grey shoots represent experimentally manipulated shoots that were supported by wooden stakes. B) Warming chambers and shelter cages installed around propped and mat krummholz shoots. C) Radiation shields with temperature loggers to measure air temperature and wax cylinders to measure winter wind effects (Maher et al. 2020).

(*Nucifraga colombiana* Wilson; Bruederle et al. 1998). Thus, krummholz *P. albicaulis* represent potential trees, making this an appropriate species for investigating environmental influences along treeline ecotones. For our purposes, we defined krummholz as matted, shrub-like *P. albicaulis* that are wider (≥ 1 m across) than they are tall.

To determine rates of establishment of natural upright stems from krummholz in the Tobacco Root Mountains, we sampled 45 upright stems at 9 sites. At each site, 5 stems were selected by standing at the highest accessible point above the local krummholz limit and randomly choosing a downslope compass direction. We sampled the first upright stem we encountered in that direction and its 4 nearest neighbors. We assumed that the current krummholz mat represents the average snow depth over the lifespan of each sampled stem. We measured each stem's vertical height above the krummholz mat and took a cross-section where it intersected the krummholz mat. In the laboratory, we counted rings to the pith on each stem cross-section and verified our counts by visually crossdating (Stokes and Smiley 1968) using a pattern of frost-damaged rings.

To test hypotheses about environmental influences limiting the krummholz growth form, we used a replicated (n = 10) factorial experiment with shoot position (upright vs. mat), late summer warming (warmed vs. ambient), shelter from winter damage (sheltered vs. exposed), and elevation position (lowest local krummholz vs. highest local krummholz) as factors and shoot mortality as the response (Fig. 2). We simulated upright shoot growth by securing naturally prostrate krummholz shoots to wooden stakes to place them ~0.5 m above the top of the krummholz mats (Fig 1.). We paired these "propped" shoots with shoots within the mat that we also secured to short wooden stakes (stakes located entirely within the mat). We used clear polycarbonate warming chambers and sturdy fine mesh cages for summer warming and winter shelter, respectively (figure 1). At each site, we installed treatments near both local high- and low-elevation limits of krummholz. Each factor was crossed such that propped/mat shoot pairs were either warmed in the summer and protected in the winter, only warmed in the summer, only protected in the winter, or neither warmed nor protected (figure 1). The experiment was replicated at 10 treeline sites (sites are blocks; n = 10). Low-elevation replicates were downslope from the higher ones. Treatments were randomly assigned to four krummholz at each low- and high-elevation location. Warming chambers were installed between 17 July to 10 August 2015. These were exchanged for shelter cages between 9 October and 15 October 2015. The experiment was recovered between 23 June and 4 July 2016.

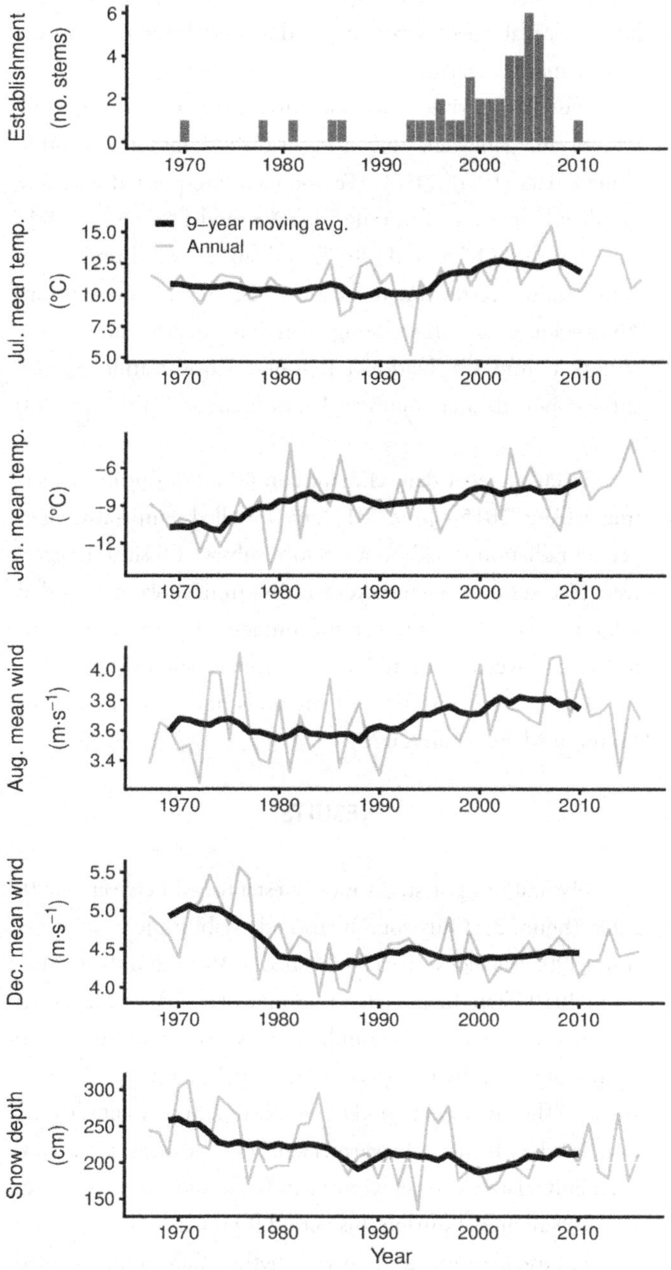

Figure 2. Number of naturally occurring emergent stems establishing each year in the treeline ecotone and associated climatic conditions during the period 1967-2016 (n = 45). Establishment represents when stem heights reached the top of the current krummholz mat. Climate variables are displayed as 9-year moving-window averages (-2 years to +6 years for each year of establishment; thick black lines) and as annual values (thin grey lines). Significant (P ≤ 0.001 in these cases) Spearman rank correlations of stem establishment with 9-year climate averages are as follows: July temp = 0.59, January temp = 0.58, August wind speed = 0.63, December wind speed = -0.23, Snow depth = -0.56 (Maher et al. 2020).

We recorded shoots as dead (in 2016) if the shoot tip was broken off or if all needles were stripped off and the apical bud was brown and desiccated.

We characterized environmental conditions during our natural emergent stem study using available long-term monthly climate data (1970-2016). We obtained interpolated monthly gridded temperature from the TopoWx model (1/120° or ~800 m resolution; Oyler et al. 2014) and surface wind time series data from the TerraClimate dataset (1/24° or ~4 km resolution; Abatzoglou et al. 2018). Long-term snow depth records were obtained from US National Resource Conservation Service snow course data at Branham Lakes (marker 11D14 at 2700 m asl).

To characterize daily climatic conditions during our experiment (Aug. 2015 - June 2016), we installed temperature loggers in radiation shields at a random subset of 5 sites. Loggers were placed on wooden stakes within krummholz mats and at ~0.5 m above the mat to capture ambient air temperature differences between mat interiors and propped shoots (figure 1).

Complete description of methods and analyses can be found in Maher et al. (2020).

RESULTS

Natural upright stems mostly established between 2000-2008 (figure 2). Only four stems had establishment dates before 1990. The oldest stem established in 1978, and the youngest in 2010. Stem heights in our sample of stems were less than or equal to 1 m above krummholz mats, except for one 1.2 m stem that established in 1998 (range = 0.2 to 1.2 m, median = 0.5 m). The observed period of establishment was significantly associated with relatively warmer summer and winter temperatures and relatively lower winter winds and snowpack. Summer winds were higher during this period (figure 2).

Exposed, propped shoots—whether they were warmed or not and across both elevations—clearly experienced higher mortality (10-50%) than did sheltered propped shoots and mat shoots (figure 3). We observed no mortality in sheltered shoots (propped or mat locations) at high-elevation and only one (10%) mat shoot (which was in a shelter cage) died at low elevation. Summer warming had no noticeable effect on this relationship at high-elevation sites. These effects (main effects and interactions) were not statistically detectable at the $P < 0.05$ level. However, the shelter by shoot position interaction was significant at the $P < 0.1$ level. Air temperature differences between above mat and withing mat environments were more pronounced during summer/fall and spring than during winter.

Complete results can be found in Maher et al. (2020).

DISCUSSION

Our findings clearly indicate the importance of winter damage in preventing upright shoot survival and growth in krummholz, although our inferences about the effects of warming are more limited. While warming chambers did increase the air temperatures shoots were exposed to, chambers

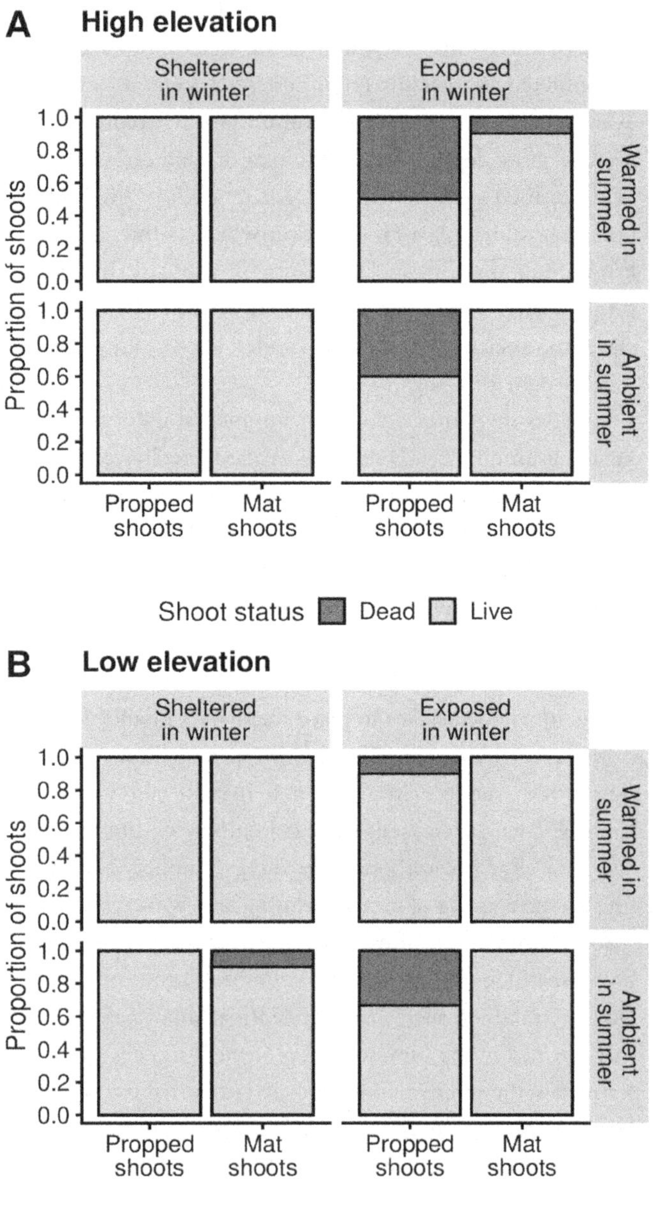

Figure 3. Post-winter survival (proportion living of all shoots; light grey) and mortality (proportion dead of all shoots; dark grey) of experimentally manipulated krummholz shoots in June/July 2016 (n = 10 sites). A) shows treatments at high-elevation, B) shows treatments at low-elevation (Maher et al. 2020).

were installed relatively late in the growing season and may not have strongly affected growth or development. Still, at low elevations there seems to have been a reduction in mortality in the exposed propped shoots that were also warmed. This could reflect greater cuticle development from warmer temperatures and thus greater resistance to winter damage (Tranquillini 1979). On the other hand, many shoots died as a result of mechanical breakage of stem tips (the top 5-10 cm was torn off). Intense physical force like this probably won't be overcome by cuticle development or a year's height increment. Dead shoots that were not broken appeared desiccated—a well-documented cause of death in conifer needles due to cuticle loss from winter wind exposure (Hadley & Smith 1986, 1989). We reason that winter damage is a first-level mechanism limiting upright growth in krummholz, with the effects of temperature on growth rates important, but subordinate.

Alpine treeline ecotones are complex ecological boundaries that are unlikely to be explained by a single factor alone (Sullivan et al. 2015; Cansler et al. 2018). Our findings of the importance of a factor other than summer temperature, winter damage in this case, calls into question the usefulness of alpine treelines as bellwethers of the effects of warming. At treelines where abrasion by wind-driven snow is possible, a warming climate may not directly result in treeline advance, unless warming also causes changes that reduce the risk of damage in winter.

ACKNOWLEDGEMENTS

This project was funded by a McIntire-Stennis Cooperative Forestry Program grant [project accession no. 225109] from the USDA National Institute of Food and Agriculture and by a Research and Creative Scholarship from the Associated Students of the University of Montana. Jessie Bunker, Megan Keville, Enzo Martelli, Mike Fazekas, Danica Born-Ropp, Aeriel Lavoie, Samuel Scott, and Pilar Mena assisted with field and lab work. Drs. Connie Millar, Anna Sala, Robert Keane, and David Affleck provided input on early manuscript drafts. Finally, we thank two reviewers for detailed feedback that helped improve the manuscript.

LITERATURE CITED

Abatzoglou JT, SZ Dobrowski, SA Parks, and KC Hegewisch. 2018. TerraClimate, a high-resolution global dataset of monthly climate and climatic water balance from 1958-2015. Sci Data. 5:1–12. doi.org/10.1038/sdata.2017.191.

Bruederle LP, DF Tomback, KK Kelly, and RC Hardwick. 1998. Population genetic structure in a bird-dispersed pine, Pinus albicaulis (Pinaceae). Can J Bot Can Bot. 76(1):83–90.

Cansler CA, D McKenzie, and CB Halpern. 2018. Fire enhances the complexity of forest structure in alpine treeline ecotones: Ecosphere. 9(2). doi.org/10.1002/ecs2.2091.

Hadley JL, and WK Smith. 1983. Influence of Wind Exposure on Needle Desiccation and Mortality for Timberline Conifers in Wyoming, U.S.A. Arct Alp Res. 15(1):127. doi.org/10.2307/1550988.

Hadley JL, and WK Smith. 1986. Wind Effects on Needles of Timberline Conifers: Seasonal Influence on Mortality. Ecology. 67(1):12–19. doi.org/10.2307/1938498.

Hadley JL, and WK Smith. 1989. Wind Erosion of Leaf Surface Wax in Alpine Timberline Conifers. Arct Alp Res. 21(4):392–398. www.tandfonline.com/doi/abs/10.1080/00040851.1989.12002752.

Hoch G, and C Körner. 2003. The carbon charging of pines at the climatic treeline: a global comparison. Oecologia. 135(1):10–21. doi.org/10.1007/s00442-002-1154-7.

Körner, C. 1998. A re-assessment of high-elevation treeline positions and their explanation. Oecologia. 115:445–459. doi.org/10.1007/s004420050540.

Körner C, and J Paulsen. 2004. A world-wide study of high altitude treeline temperatures. J Biogeogr. 31(5):713–732. http://doi.wiley.com/10.1111/j.1365-2699.2003.01043.x.

Maher CT, CR Nelson, and AJ Larson. 2020. Winter damage is more important than summer temperature for maintaining the krummholz growth form above alpine treeline. J Ecol. 108(3):1074–1087. doi.org/10.1111/1365-2745.13315.

Oyler JW, A Ballantyne, K Jencso, M Sweet, and SW Running. 2014. Creating a topoclimatic daily air temperature dataset for the conterminous United States using homogenized station data and remotely sensed land skin temperature. Int J Climatol. doi.org/10.1002/joc.4127.

Rogers DL, CI Millar, and RD Westfall. 1999. Fine-Scale Genetic Structure of Whitebark Pine (Pinus albicaulis): Associations with Watershed and Growth Form. Evolution

(N Y). 53(1):74–90. doi.org/10.1111/j.1558-5646.1999.tb05334.x.

Stokes M, and TL Smiley. 1968. An introduction to tree–ring dating. Chicago: University of Chicago Press.

Sullivan PF, SBZ Ellison, RW McNown, AH Brownlee, and B Sveinbjörnsson. 2015. Evidence of soil nutrient availability as the proximate constraint on growth of treeline trees in northwest Alaska. Ecology. 96(3):716–727. doi.org/10.1890/14-0626.1.

Tranquillini, W. 1979. Physiological Ecology of the Alpine Timberline. Berlin, Heidelberg: Springer (Ecological Studies). doi.org/10.1007/978-3-642-67107-4. Accessed 26 June 2014.

Long-term Assessment of the Efficacy of Prescribed Burning and Mechanical Cutting for Restoration of Whitebark Pine

Enzo Martelli, <enzo.martellimoya@umontana.edu>, Department of Ecosystem and Conservation Sciences, W.A. Franke College of Forestry and Conservation, University of Montana, 32 Campus Drive, Missoula, MT; **Robert E. Keane**, US Forest Service Rocky Mountain Research Station Missoula Fire Sciences Lab; **Cara R. Nelson**, Department of Ecosystem and Conservation Sciences, W.A. Franke College of Forestry and Conservation, University of Montana, 32 Campus Drive, Missoula, MT.

ABSTRACT

Whitebark pine is declining across most of its range due to the combined effects of an invasive pathogen, episodic native insect outbreaks, climate-change-induced increases in wildfire frequency, and successional replacement due to fire suppression. Prescribed burning and mechanical cuttings have been proposed as primary strategies for restoration. However, there is limited information on the efficacy and effects of these treatments, and projects that have been evaluated show highly variable results, suggesting that monitoring is a priority need. We used a 15-year, replicated before-after-control-impact study to assess the effects of prescribed burning and mechanical cuttings on mortality, seedling density change, and basal area change for competing conifers in the northern Rocky Mountains of the United States. Whitebark pine mortality (plot percentage) was about 20% higher for prescribed burning alone and prescribed burning and mechanical cutting than control plots. Seedling density change (ind./ha) was 1% lower and 14% higher than control plots for prescribed burning alone and prescribed burning and mechanical cutting, respectively. Basal area change (m^2/ha) of *Abies lasiocarpa* was 61% and 98% lower than control plots for prescribed burning alone and prescribed burning and mechanical cutting, respectively. Since the implementation of this experiment in the mid-90s, whitebark pine communities have shown a significant increment in mortality of mature trees and reduced natural regeneration in addition to the treatments effect. These results indicate that active management is an effective strategy to mitigate the effect of current threats, where trade-off's include promoting natural regeneration at the cost of greater tree mortality.

INTRODUCTION

Whitebark pine (*Pinus albicaulis*), an ecologically important tree species of high-elevation ecosystems of western North America, is declining across most of its range due to the combined effects of an invasive pathogen (white pine blister rust; *Cronartium ribicola*), episodic native insect (mountain pine beetle; *Dendroctonus ponderosae*) outbreaks, climate-change-induced increases in wildfire frequency and severity (Keane and Arno 1993; Westerling et al. 2006) as well as successional replacement as a result of a century of fire suppression in the mesic part of its distribution. Concern over these threats has led to a petition to list whitebark pine as a threatened species under the US Endangered Species Act (FWS 2021), the designation as a species at-risk under the Canadian Endangered Species Act (NRDC 2008), and an increase in research and management activities. Despite a growing understanding and widespread agreement on the need for conservation and restoration (Keane et al. 2012), there is little information on the species' response to management actions. Specifically, the extent to which prescribed fire and mechanical cuttings are effective for meeting restoration objectives. To assess long term effects of prescribed burning and mechanical cuttings on whitebark pine forests, we analyzed data collected over 15 years in the Bitterroot Mountains, Montana, USA.

Prescribed burning and mechanical cutting are included as key activities in the Range-Wide Restoration Strategy for Whitebark Pine (Keane et al. 2012) and in the National Whitebark Pine Restoration Plan. The goals of these treatments are to promote diverse age-class structure, reduce competing vegetation, increase the vigor of mature whitebark trees, encourage natural regeneration, and reduce mountain pine beetle hazard (Keane et al. 2012). Nonetheless, there is still uncertainty on the ecological viability of these restoration activities and their potential outcomes (Maher et al. 2018; Keane 2018; Tomback et al *in review*). Currently, few mechanical cuttings, and even fewer prescribed burns, have been monitored for their ecological effects, and for the ones that have, findings are mixed, suggesting that additional monitoring is a priority need. Also, because effects of treatments on seedling recruitment are notoriously difficult to study due to high spatial and temporal variability in high-elevation seedling establishment driven by decadal-scale oscillations in climate (Youngblut and Luckman 2013), determining regeneration responses requires sampling over long timeframes with standardized methods capable of capturing an effect. Here we took advantage of the only existing long-term before-after-control-impact (BACI) replicated study, that we are aware of, on the efficacy and ecological effects of mechanical cuttings and prescribed burns in whitebark pine forests to ask questions about treatment efficacy and effects. Our specific research question is: What is the magnitude of difference among treatments (prescribed burn, prescribed burn and mechanical cutting, and untreated control) in whitebark pine mortality, abundance, and regeneration, as well as in the abundance of competitor conifer species (*Abies lasiocarpa, Picea engelmannii, Pinus contorta*), over a 15-year period?

MATERIALS AND METHODS

Experimental design: This study is part of an on-going long-term monitoring project, "Restoring Whitebark Pine Communities" (Keane and Parsons 2010b), that was designed to test the effects of prescribed burning and mechanical cutting on whitebark pine ecosystems. The study included a combination of experimental silvicultural treatments and prescribed burning with or without fuel bed enhancement. The study was implemented at five sites (Bear Overlook, Beaver Ridge, Coyote Meadows, Musgrove, and Smith Creek) in the northern Rocky Mountains of the United States. We analyzed data from just one of the five sites, Bear Overlook, because it was the only site that met the following experimental conditions: (1) had treated areas and controls sampled in the 15th year after treatment implementation, and (2) had been unaffected by post-treatment wildfire events.

At the Bear Overlook site, ten 0.04 ha plots were installed at each of three treatment units (prescribed burning alone, prescribed burning and mechanical cutting, and untreated control). In 1996, vegetation was sampled; live trees above 12 cm were tagged and species, DBH, tree height, height to crown base, and health were recorded. The same measurements were also recorded on all live trees less than 12 cm DBH (in 2.5 cm DBH classes) and greater than 1.37 m tall (saplings). *P. albicaulis* seedlings (trees less than 1.37 m tall) were counted within a 125 m^2 circular plot nested within each 0.04 ha plot. In 1999, a low-intensity underburn was implemented in both units that included prescribed burning. In one of these units, prior to the burn, all trees from competing species were harvested. Vegetation was sampled again the year after treatment implementation, and every 5 years after that. For the purposes of this study, we analyzed the 15-year post-treatment data to test for responses since the last assessment (Keane and Parsons 2010b).

To assess the effect of prescribed burning alone and in combination with mechanical cutting on whitebark pine mortality (proportion of live trees sampled pre-treatment that were dead by the 15th year post-treatment), we calculated abundance (relative change in basal area pre-treatment to last post-treatment sample year), regeneration (relative change in seedling density pre-treatment to last post-treatment sample year), and abundance (change in basal area pre-treatment to last post-treatment sample year) of competing conifers, and used non-parametric Kruskal-Wallis tests to determine statistical significance for each response variable.

RESULTS

Treated stands had 80% and 78% whitebark pine mortality (prescribed burning alone and prescribed burning and mechanical cutting, respectively), about 20% higher than the mortality in the untreated stand. However, this trend was not statistically significant ($\chi^2 = 2.34$, df = 2, $\rho = 0.31$, figure 1A). Over the 15-year period in the control stand, *A. lasiocarpa*, *P. contorta*, and *P. engelmannii* increased in basal area by 77% (± 28 SE), 7% (± 18 SE), and 10% (± 4 SE), respectively,

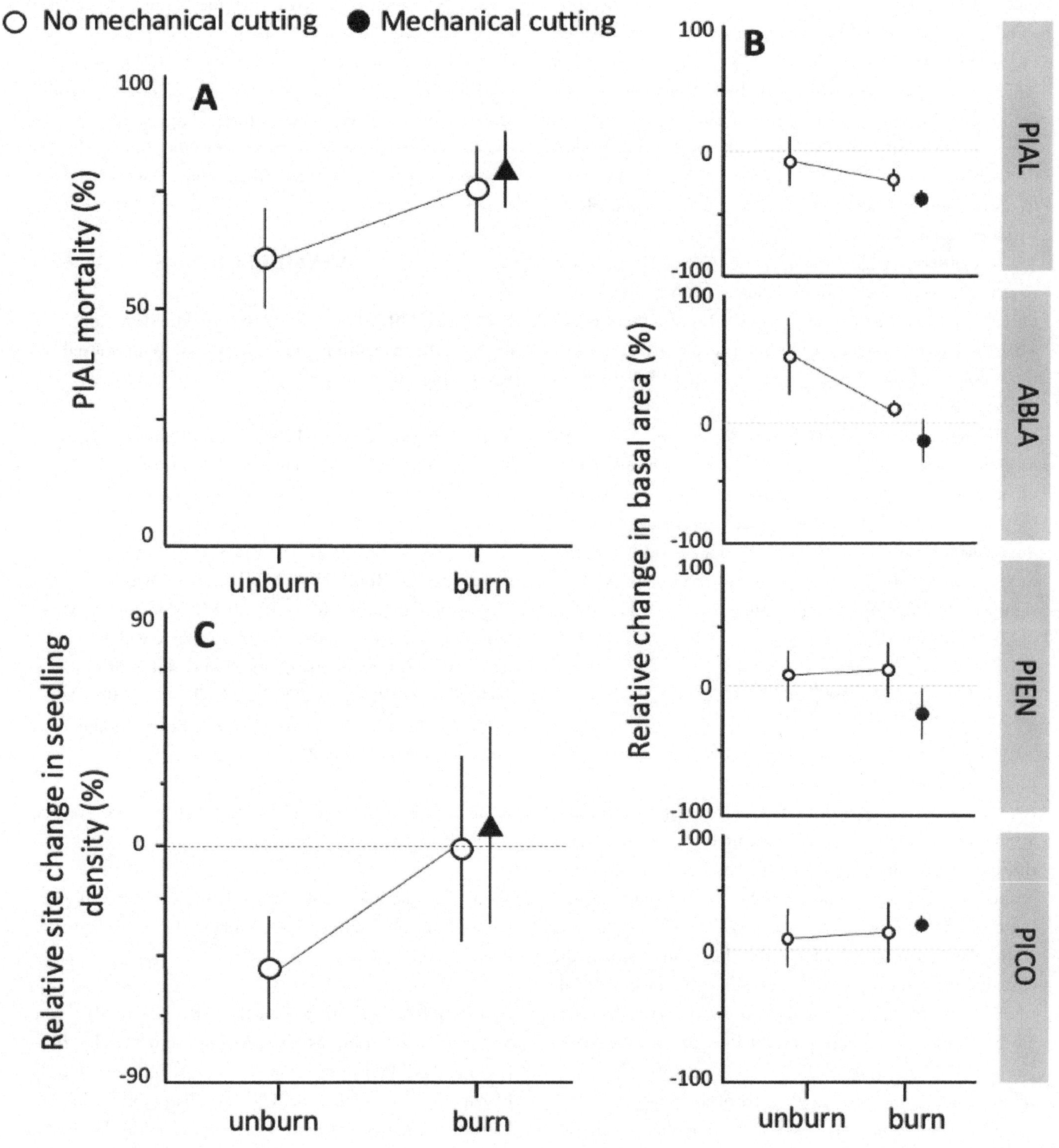

Figure 1. PEffect of prescribed burning (x-axis) and mechanical cutting (closed symbols, harvested; open symbols, not harvested) on (A) mean (± SE) adult whitebark pine tree mortality (% over the 15 year period); (B) mean (± SE) relative change (% change pre to post treatment) in live tree basal area (m^2/ha) of *Pinus albicaulis* and three shade-tolerant conifer species; and (C) mean (± SE) relative change (% change pre to post treatment) in whitebark pine seedling density. PIAL = *Pinus albicaulis*, ABLA = *Abies lasiocarpa*, PIEN = *Picea engelmannii*, PICO = *Pinus contorta*.

while *P. albicaulis* decreased by 8% (± 20 SE) from pre-treatment levels. In comparison, in the prescribed burning alone stand, *A. lasiocarpa*, *P. contorta*, and *P. engelmannii* increased in basal area by 16% (± 7 SE), 13% (± 21 SE), and 15% (± 23 SE), respectively, while whitebark decreased by 27% (± 13 SE). In the prescribed burning and mechanical cutting treatment, basal area of *A. lasiocarpa* and *P. engelmannii* declined by 21% (± 25 SE), and 23% (± 21 SE), respectively, and *P. contorta* increased by 18% (± 8 SE), while whitebark decreased by 45% (± 6.1 SE).

The observed changes in basal area (pre to 15 years post treatment) were significantly different among treatments only for *A. lasiocarpa* ($\chi^2 = 10.29$, df = 2, $\rho = 0.0058$ figure 1B). There were no significant differences among treatments for *P. contorta* ($\rho = 0.27$), *P. engelmannii* ($\rho = 0.33$), or *P. albicaulis* ($\rho = 0.31$).

In the control, seedling density declined by 37% (± 19 SE). In comparison, in the prescribed burning only treatment, seedling density declined by only 1% (± 31 SE) and, in the prescribed burning and mechanical cutting treatment, it increased by 14% (± 37SE). However, the change in seedling density pre to 15-years post treatment was not significantly different among treatments ($\chi^2 = 2.34$, df = 2, $\rho = 0.55$) (figure 1C).

DISCUSSION

Given increasing concern about declines in whitebark pine across its range (Loehman et al. 2018), there is a corresponding need to examine effects of restoration and management treatments being implemented to mitigate current threats. Our results suggest that the combined effect of prescribed burning and mechanical cutting reduced competing conifers (*A. lasiocarpa* and *P. engelmannii*) and released whitebark pine seedlings. However, this result came at the cost of higher mortality of potentially cone-bearing adult trees. This setback supports other observations that fire may adversely affect whitebark pine (Nelson and Keville 2018, Keane 2018), even at relatively low levels of bole scorch (Hood et al. 2008) over the 15 years of this study.

One limitation that could affect inference from our study is lack of replication. We assessed only one site. Although the original study included five sites, four were disturbed by post-treatment wildfire. Furthermore, only two of the original sites had data collected from both treated areas and controls by the 15th year after treatment implementation. The high probability of disturbance and other chance events impacting project sites, coupled with high variability in both treatment prescriptions and their effects (Maher et al. 2018), suggest the need to include a much larger number of replicates in assessments of treatment effects. Given that, the data needs for inference may be too large for any one administrative unit or research project alone. Consideration should be given to developing a large-scale long-term monitoring network to improve understanding of the efficacy and effects of restoration treatments in whitebark pine ecosystems.

LITERATURE CITED

Hood, SM, DR Cluck, SL Smith, and KC Ryan. 2008. Using bark char codes to predict post-fire cambium mortality. Fire Ecology 4(1):57-73.

Keane, RE, and SF Arno. 1993. Rapid Decline of Whitebark Pine in Western Montana: Evidence from 20-Year Remeasurements. Western Journal of Applied Forestry 8: 44-47.

Keane, RE, DF Tomback, CA Aubry, AD Bower, EM Campbell, CL Cripps, MB Jenkins, MF Mahalovich, M Manning, and ST McKinney. 2012. A range-wide restoration strategy for whitebark pine (Pinus albicaulis), Gen. Tech. Rep. RMRS-GTR-279. Fort Collins, CO: US Department of Agriculture, Forest Service, Rocky Mountain Research Station 108 p. 279.

Keane, RE 2018. Managing Wildfire for Whitebark Pine Ecosystem Restoration in western North America. Forests 9: 22.

Keane, RE, and RA Parsons. 2010. Restoring whitebark pine forests of the northern Rocky Mountains, USA. Ecological Restoration 28: 56-70.

Loehman, RA, BJ Bentz, GA DeNitto, RE Keane, ME Manning, JP Duncan, JM Egan, MB Jackson, S Kegley, IB Lockman, DE Pearson, JA Powell, S Shelly, BE Steed, and PJ Zambino. 2018. Effects of Climate Change on Ecological Disturbance in the Northern Rockies. in Halofsky, JE and Peterson, DL (eds.). Climate Change and Rocky Mountain Ecosystems (Springer: Dordrecht).

Maher, CT, CR Nelson, AJ Larson, and A Sala. 2018. Ecological effects and effectiveness of silvicultural restoration treatments in whitebark pine forests. Forest Ecology and Management 429: 534-48.

Nelson, CR, and MP Keville. 2018. Effects of Prescribed Burning on Whitebark Pine. Stanley, Idaho: Whitebark Pine Ecosystems Foundation.

NRDC. 2008. A Petition to List the Whitebark Pine, Pinus albicaulis, as an Endangered Species Under the Endangered Species Act.

Tomback, DF, RE Keane, AW Schoettle, RA Sniezko, MM Jenkins, MF Mahalovich, CR Nelson, E Jungck, S Kegley, E Pansing, E Guiberson, A Bower, C DeMastus, M Murray, R Mangold, J Shamhart, E, Sprague, D and Gwaze. In review. Current and best management practices for the restoration of whitebark pine (Pinus albicaulis), a threatened high-elevation western North American forest tree. Forest Ecology and Management.

Westerling, AL, HG Hidalgo, DR Cayan, and TW Swetnam. 2006. Warming and earlier spring increase western U.S. forest wildfire activity. Science 313: 940-3.

FWS. 2020. Endangered and Threatened Wildlife and Plants; Threatened Species Status for Pinus albicaulis (Whitebark Pine) With Section 4(d) Rule. https://www.federalregister.gov/d/2020-25331. Accessed 25 November 2021.

Youngblut, DK, and BH Luckman. 2013. Evaluating the temperature sensitivity of radial growth patterns from whitebark pine in the western Canadian Cordillera. Dendrochronologia 31: 16-28.

How to Retain Whitebark Pine in Timber Harvests

Michael P. Murray, <michael.murray@gov.bc.ca>, British Columbia Forest Service, Nelson, BC; **Jenny Berg**, <vedaworks@gmail.com>, University of Victoria, Victoria, BC; **Dave Huggard**, <dhuggard@telus.net>, Apophenia Consulting, Vancouver, BC.

ABSTRACT

In British Columbia, whitebark pine is a component of harvested forests, yet knowledge of post-harvest survivorship and factors that promote successful retention is lacking. Our objectives were to describe the temporal attrition of retained mature whitebark pine trees and to identify factors that likely influence survivorship during the critical initial post-harvest period. We assessed five separate harvest units in southeastern British Columbia. We found that retained trees experienced high annual mortality rates (3%-16%) across harvest sites during the initial five-year post-harvest period. After eight years post-harvest, mortality rates drastically declined. The preponderance of fallen stems oriented towards the northeast suggests that storm system events arriving from the Pacific Ocean are the most significant drivers of blowdown. We estimate that survivorship is positively associated with shorter tree heights and longer crown lengths, a lack of disease cankers, a greater presence of rodent wounding, and higher numbers of surrounding retained trees. Slope and aspect had very minor influence. As these trees are an endangered species, harvest operations should be practiced cautiously in associated forests.

Keywords: blowdown, fire, harvest, *Pinus albicaulis*, silviculture, whitebark pine, wind

INTRODUCTION

Whitebark pine trees are widely distributed among subalpine mixed-conifer forests in southern British Columbia. The most abundant associated tree species are Engelmann spruce (*Picea engelmannii*) and subalpine fir (*Abies lasiocarpa*). A gradual increase in harvest acreage above 1800 m elevation began in 2008 in the Kootenay-Boundary Region. The long-term retention of mature whitebark pine trees can ensure that ecological values are better protected.

Before this study, survivorship of whitebark pine retained within commercial harvests had not been examined. We investigated the fate of residual trees to infer some preliminary recommendations. Our objectives were to describe the temporal attrition of retained mature whitebark pine trees and to identify factors that likely promote survivorship during the critical initial post-harvest period.

METHODS

Field sampling was conducted at five separate harvest sites during the summer of 2018. All five sites are located in the Kootenay-Boundary region of southeastern British Columbia. They are considered variable retention cuts and represent the only harvest method deployed to date in the region's whitebark pine stands. Because harvest years differed among the sites, our field sampling captured a range of post-harvest intervals representing 6–17 years. Based on the majority of whitebark pine stands in the region, our study sites are representative of the most common mix of tree species, elevation range, tree ages, and habitat.

At each harvest site a census of standing and downed mature whitebark pine greater than 17 cm dbh (diameter at breast height) was conducted. Trees near the perimeter of the harvest units were excluded if the tree height was greater than the distance to the nearest forest edge. The location of every tree was recorded with a high-precision (GNSS) global positioning device. Every tree was assessed based on the following parameters: height, diameter (dbh), distance and azimuth to nearest forest edge, height to live crown, and percentage of live crown. Slope (%) and aspect (deg) were measured at each tree. For every surveyed tree we tallied the number of mature neighbor trees within a distance equal to or less than the survey tree's height. Each neighbor tree was noted according to status (live, snag, down). All trees were examined for forest health agents.

For dendrochronological analysis, increment cores were collected from all live and dead retained trees using a 4-mm Haglof increment borer taken at approximately 1.3 m above the ground (dbh). Those samples with exceptionally condensed rings were measured with a Velmex uniSlide digitally encoded traversing table at a precision of 0.01 mm. The remaining cores were digitally scanned at a 2400 dpi resolution with an HP flatbed scanner. Digital images were imported into CooRecorder measuring software and exported as ring width files with CDendro software package (Larsson 2014). To ensure that the appropriate calendar date was assigned to each measured ring we used the program COFECHA to aid in accurately cross-dating all increment cores (Holmes et al. 1986).

To identify potential factors affecting survivorship, we evaluated a set of models using Akaike's "An Information Criterion" (AIC) (Burnham and Anderson 1998). Biophysical measurements were examined as potential factors to predict the survival of retained whitebark pine trees. To reduce the number of variables included in the analysis, correlated variables within each set were screened using AIC model selection. The best subset of variables in each set was used in the final analysis. The final analysis used AIC model selection with 16 models, representing all combinations of including or excluding the selected variables in each of the four sets. All models were general linear mixed-effects models. The analyses were conducted separately for windthrown versus live trees, and for all dead trees (windthrown and standing dead) versus live trees.

RESULTS

We analyzed a total of 197 dead trees and 134 live trees. Mortality rates were highest immediately following harvest (figure 1). At Lavington (LV) operators reported that most retained trees were blown over during a single powerful storm as they were completing harvest. A negative exponential trend characterized three harvest sites, where initial steep declines became increasingly moderated over time. By nine years post-harvest, mortality ceased at all but a single harvest site.

The probability of mortality of retained whitebark pine trees is best explained by a combination of tree characteristics, slope/aspect, and the number of surrounding retained

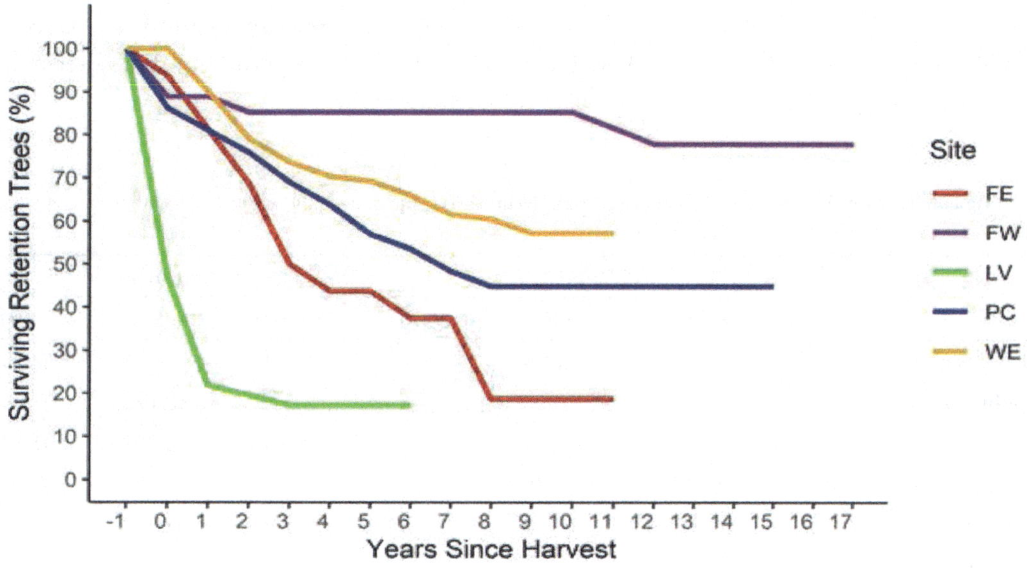

Figure 1. Post-harvest attrition of retained whitebark pine according to harvest site (FE: Findlay East; FW: Findlay West; LV: Lavington; PC: Paturages; WE: West Elk) (Murray et al. 2021).

trees (table 1). We found a strong increase in survivorship, with greater tree crown length accompanied by decreasing tree height. Thus, the probability of post-harvest mortality was higher for taller trees with shorter crowns and lower for shorter trees with long crowns. In examining the importance of neighbor trees, a survivorship probability greater than 50% required a minimum of 7.5 retained neighbor trees within tree height radial distance. For trees that did not survive, we found the vast majority of downed stems oriented in a northeasterly direction from root collar to crown indicating the strongest winds experienced at the sites arrived from southwesterly directions. Interestingly, there were opposite effects depending on the tree lesion type (cankers vs. rodent wounding). Any rodent damage indicated higher survivorship. With one or more blister rust cankers, there would be less than a 50% chance of survival.

DISCUSSION

We found elevated mortality rates occurred during the initial five-year post-harvest period. Due to a paucity of research on whitebark silviculture, it is unclear how whitebark pine compares with the capabilities of other tree species. Alternate species were not retained in our study sites, except at Lavington with very low numbers. The results suggest that most trees fell during storm conditions. We suggest that winter storms and approaching fronts of coastal low-pressure systems are the most significant drivers of blowdown for whitebark pine stands in the southern interior region. For at least one harvest site (Lavington), a majority of trees were blown over while alive.

Although cankered mature trees can survive for decades, if *Cronartium ribicola* remains in the host, chronic stress may interfere with physiological mechanisms that contribute to the tree's ability to withstand wind. Contrary to expectations, we found higher survival in trees damaged by rodents. Rodent damage may therefore indicate healthier trees that can adapt more quickly to post-harvest exposure.

The probability of survival lessened for trees with shorter crown lengths and greater heights. In general, trees that grew in denser stands with resultingly shorter crowns may be less adapted to resisting windthrow when they are exposed at harvest. Furthermore, the top-heaviness seems to make these trees more vulnerable. Our results are consistent with the vast majority of retention studies, indicating that higher retention levels favor positive survivorship rates (e.g., Busby et al. 2011; Montoro Girona et al. 2019; Moussaoui et al. 2020; Rosenvald et al. 2008). There are likely additional factors that favor retention survivorship that we did not examine, which may include pre-harvest stem density, soil (texture, depth, moisture), and rooting structure.

After completion of our study, a fire impacted the Lavington harvest site ("Doctor Creek Wildfire") in late summer of 2020 (figure 2). A survey was conducted in 2021 to determine the post-fire status of the 16 trees that were alive when our research sampling completed in 2018. Of these 16 trees, only one tree was alive (only 10% of foliage was green). Five trees had blown over between 2018-2020 and were consumed in the fire. Overall, 10 trees appeared to be directly killed by the fire. Bole scorch height varied from 25-80% of total height. Of note, about half of the retained western larch (*Larix occidentalis*) trees were alive. The fire resulted in a stand-replacing burn in the surrounding forest. Within the harvest unit, most coarse woody debris was consumed as well as a substantial proportion of duff cover indicating a high intensity event.

Table 1. Logit-scale coefficients for the best model for all dead trees (windthrow and snags) and windthrow only (Murray et al. 2021).

	All Dead Trees			Windthrow		
	Estimate	SE	*p*	Estimate	SE	*p*
Intercept	-0.918	1.154	0.427	-0.792	1.353	0.558
Nfacing	-	-	-	2.399	1.173	0.041
Efacing	-	-	-	-1.907	1.268	0.132
Sqrt(Surrounding Live Down)	-0.447	0.179	0.012	-0.524	0.188	0.005
nCankers	0.141	0.088	0.109	0.249	0.096	0.010
Rodent	-1.760	0.397	<0.001	-2.312	0.468	<0.001
Ht	0.281	0.071	<0.001	0.2460	0.079	0.002
CrownLength	-0.342	0.069	<0.001	-0.3351	0.074	<0.001

Figure 2. Post-harvest retained whitebark pine trees that were killed by the 2020 Doctor Creek Wildfire at the Lavington harvest site near Canal Flats, BC.

MANAGEMENT IMPLICATIONS

For southeast British Columbia and the adjacent Kootenai Region of the USA, we recommend harvest practitioners carefully retain whitebark pine. To increase likelihood of survival, retaining a minimum of eight neighboring trees (within the target tree's height radius) will substantially reduce risk. Choose trees with longer crown lengths and lower frequencies of disease cankers. Trees above average height are at higher risk of becoming windfall. We further recommend that harvest planners lay out ovate patches of retention oriented on a southwest-to-northeast azimuth to reduce hazards from windstorms. Harvesters should consider moving any wood debris away from retained stems. During fire events, we suggest that retained trees be protected by clearing surface fuels away from their driplines, wrapping tree boles with resistant material, and conducting spot suppression (Keane 2018; Murray 2007). All healthy cone-bearing trees are potentially disease resistant, thus represent a life link to the species' future.

ACKNOWLEDGEMENTS

Many thanks to Katy Fraser, Marnie Duthie-Holt, Jesse Ellis, Roger Powell, Peter Vandenbergh, Jodi Axelson, Tara DeCourcy, Amanda Asay, Jodie Krakowski, Jamie Hibberson, David Karassowitsch, Stephanie Keightley, Randy Moody, Paul Mozin, and Stephen Mitchell.

LITERATURE CITED

Burnham KP, and DR Anderson. 1998. Practical use of the information-theoretic approach. In Model selection and inference: A practical information-theoretic approach, edited by KP Burnham and DR Anderson, 75-117. New York, NY: Springer. https://doi.org/10.1007/978-1-4757-2917-7_3.

Busby PE, P Adler, TL Warren, and FJ Swanson. 2011. Fates of live trees retained in forest cutting units, western

Cascade Range, Oregon. Canadian Journal of Forest Research. https://doi.org/10.1139/xO6-140.

Holmes RL, RK Adams, and HC Fritts. 1986. Tree-ring chronologies of western North America: California, eastern Oregon and northern Great Basin with procedures used in the chronology development work including users manuals for computer programs COFECHA and ARSTAN. Laboratory of Tree-Ring Research Archives, The University of Arizona. https://repository.arizona.edu/handle/10150/304672. Accessed 24 Mar 2021.

Keane RE. 2018. Managing wildfire for whitebark pine ecosystem restoration in western North America. Forests. 9(10):648. https://doi.org/10.3390/f9100648.

Larsson, L. 2014. CooRecorder and Cdendro programs of the CooRecorder. Cdendro package version 7. http://www.cybis.se/forfun/dendro/. Accessed 17 May 2021.

Montoro, GM, H Morin, J-M Lussier, and J-C Ruel. 2019. Post-cutting mortality following experimental silvicultural treatments in unmanaged boreal forest stands. Frontiers in Forests and Global Change 2:4. https://doi.org/10.3389/ffgc.2019.00004.

Moussaoui, L, A Leduc, MM Girona, AC Bélisle, B Lafleur, NJ Fenton, and Y Bergeron. 2020. Success factors for experimental partial harvesting in unmanaged boreal forest: 10-year stand yield results. Forests 11(11):1199. https://doi.org/10.3390/f11111199.

Murray, MP, J Berg, and DJ Huggard. 2021. Harvest retention survivorship of endangered whitebark pine trees. Forests 12(6):654. https://doi.org/10.3390/f12060654.
Murray, MP. 2007. Fire and Pacific Coast whitebark pine. In Proceedings of the conference on whitebark pine: a Pacific Coast perspective, 2006 August 27-31, Ashland, OR, R6-NR-FHP-2007-01, technical coordinators EM Goheen and RA Sniezko, 51-60. Portland, OR: U.S. Department of Agriculture, Forest Service, Pacific Northwest Region. https://www.fs.usda.gov/Internet/FSE_DOCUMENTS/fsbdev2_026092.pdf. Accessed 20 Dec 2021.

Rosenvald R, A Lõhmus, and A Kiviste. 2008. Preadaptation and spatial effects on retention-tree survival in cut areas in Estonia. Canadian Journal of Forest Research 38(10):2616-2625. https://doi/10.1139/X08-093.

Preparing for Invasion: Rust Resistance in Limber, Great Basin Bristlecone, and Rocky Mountain Bristlecone Pines

Anna W. Schoettle, <anna.schoettle@usda.gov>, USDA Forest Service, Rocky Mountain Research Station, Ft. Collins, CO; **Angelia Kegley**, USDA Forest Service, Pacific Northwest Region, Dorena Genetic Resource Center, Cottage Grove, OR; **Richard A. Sniezko**, USDA Forest Service, Rocky Mountain Region, Forest Health Protection, Lakewood, CO; **Kelly S. Burns**, Retired, USDA Forest Service, Pacific Southwest Research Station, Placerville, CA; **Detlev Vogler**, USDI Park Service, Great Sand Dunes National Park and Preserve, Mosca, CO; **Phyllis Pineda Bovin**, USDI Park Service, Great Sand Dunes National Park and Preserve, Mosca, CO; **Gretchen Baker**, USDI Park Service, Great Basin National Park, Baker, NV.

Rocky Mountain (RM) bristlecone pine (*Pinus aristata*), Great Basin (GB) bristlecone pine (*P. longaeva*), and limber pine (*P. flexilis*) are high-elevation five-needle pines threatened by the non-native pathogen Cronartium ribicola that causes the disease white pine blister rust (WPBR). The pathogen continues to spread, and the infection front is now in the southern Rocky Mountains and Great Basin. WPBR is increasing in these areas on limber pine and RM bristlecone pine as well as whitebark pine (*Pinus albicaulis*); WPBR has not yet been documented on GB bristlecone in its native range. Because many of these populations are still healthy, they offer opportunities to assess baseline frequencies of WPBR resistance traits in largely naïve populations and to make comparisons among host species (Schoettle et al. 2012). Information on genetic resistance traits and their frequencies also supports development of proactive interventions to increase the frequency of resistance in populations before pathogen invasion, to mitigate future impacts and sustain ecosystem function during pathogen naturalization (Burns et al. 2008; Schoettle and Sniezko 2007; Schoettle et al. 2019a, b).

We present results from the first extensive *C. ribicola* inoculation common garden studies to assess quantitative resistance in seedling families for limber and the two bristlecone pine species (table 1). Seedling families were grown from seed collected from stands that, at the time, had little to no WPBR infections and no WPBR-caused mortality and were therefore randomly selected from these populations. More complete analyses and interpretation of the results from these studies will be presented elsewhere. Other studies, not reported here, have been conducted to assess each of these three species for qualitative resistance to WPBR (i.e., complete resistance, major gene resistance, MGR), which was only detected in limber pine (Schoettle et al. 2014; Schoettle et al. unpublished data; Sniezko et al 2016; Vogler et al. 2006).

RM bristlecone pine, GB bristlecone pine, and limber pine are highly susceptible to WPBR in seedling screening trials though they differ in their response to inoculation (figure 1). Disease resistance was evident in some families of each species and varied among species (data not shown) resulting in differential survival after inoculation (figure 2). The RM bristlecone pine seedlings in Study B were younger at the time of inoculation than those in Study A (2 years and 3.5 years, respectively) which likely contributed to the more rapid decline in vigor and post-inoculation survival in Study B (figure 2). Variation in disease progression and rates of mortality following inoculation were observed among seedling families and source areas for each species (data not shown). These studies suggest that (1) southern Rocky Mountain sources of limber pine appear to have a very low frequency of quantitative resistance to WPBR, and (2) both bristlecone pine species have higher frequencies of quantitative resistance, which may be the highest of the North American five-needle pines. The relative susceptibility of limber pine and RM bristlecone

Figure 1. Examples of symptoms of WPBR on seedlings of RM bristlecone (left), GB bristlecone (center), and limber (right) pine observed after artificial inoculation with *C. ribicola* at DGRC (Study C).

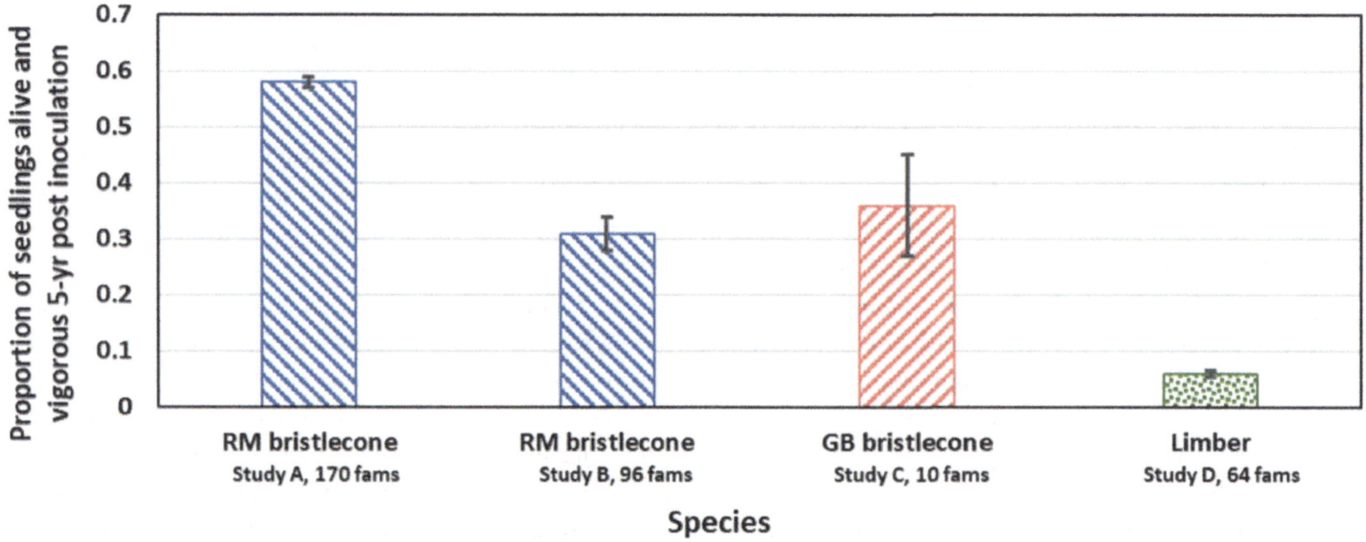

Figure 2. Preliminary results of the proportion of seedlings of each family, species and study that were alive and vigorous five years post artificial inoculation. Variation among families is shown as standard errors. MGR families of limber pine were removed from the average survival reported here. Results are from studies A, B, C (GB bristlecone only), and D (see table 1).

pine agrees with smaller, earlier trials by others (Hoff et al. 1980; Stephan 2004) though the two bristlecone species were mixed and referred to as *P. aristata* in one, and possibly both, earlier tests. As observed for quantitative resistance in other white pines, there is some evidence the resistance may be lower under higher disease pressure for RM bristlecone pine (Jacobi et al. 2018), and the future rust hazard of the environments in which these species occur under climate change may influence overall survival. These species are susceptible to WPBR suggesting that if a stand does not currently contain WPBR-diseased trees it is likely a result of ecological context, and not necessarily genetic resistance of the pines to WPBR. That is, the dry habitats and low *C. ribicola* spore availability of the Great Basin, conditions that may change in the future, likely contribute to why GB bristlecone pine is not currently diseased in the field. However, in the presence of *C. ribicola*, genetic resistance to WPBR is expected to be a significant determinant of a species' population trajectories. Field verification of WPBR resistance expression under natural conditions for RM bristlecone and limber pine families is ongoing as part of the Southern Rockies Rust Resistance Trial in southern Wyoming (Schoettle et al. 2018) and a clone bank of grafted material from the resistant limber pine seedlings has been established in Colorado.

Table 1. The first extensive artificial inoculation trials of individual-tree seed lots (families) of RM bristlecone, GB bristlecone, and limber pine for quantitative resistance to white pine blister rust. Studies are collaborations between Rocky Mountain Research Station, Rocky Mountain Region Forest Health Protection, National Park Service (Great Sand Dunes National Park & Preserve and Great Basin National Park), Pacific Northwest Region Dorena Genetic Resource Center (DGRC), and Pacific Southwest Research Station Institute of Forest Genetics (IFG). The inoculations were conducted at DGRC (Studies A, C, D, and E) and the Vogler lab at IFG (Study B) and phenotypes were assessed collaboratively. Methods are described in Schoettle et al. (2011) and Vogler et al. (2006). Study E is on-going (results not reported here).

Study	Species	# Families	Seed Sources
A	RM bristlecone	170	CO
B	RM bristlecone	96	CO
C	GB bristlecone	10	NV
C	RM bristlecone	2	CO
C	Limber	3	CO, WY
D	Limber	64	CO, WY
E	GB bristlecone	20	NV
E	Limber	20	NV

LITERATURE CITED

Burns KS, AW Schoettle, WR Jacobi, and MF Mahalovich. 2008. Options for the management of white pine blister rust in the Rocky Mountain Region. USDA Forest Service General Technical Report RMRS-GTR-206. 26 p. https://www.fs.usda.gov/treesearch/pubs/29450. Accessed 02 February 2022.

Hoff R, RT Bingham, and GI McDonald. 1980. Relative blister rust resistance of white pines. European Journal of Forest Pathology 10:307-316. https://onlinelibrary.wiley.com/doi/epdf/10.1111/j.1439-0329.1980.tb00042.x.

Jacobi WR, HSJ Kearns, A Kegley, DP Savin, R Danchok, and RA Sniezko. 2018. Comparative look at rust infection and resistance in Limber Pine (Pinus flexilis) and Rocky Mountain Bristlecone Pine (P. aristata) following artificial inoculation at three inoculum densities. In Proceedings of the IUFRO joint conference: Genetics of five-needle pines, rusts of forest trees, and Strobusphere, 15-20 June 2014, Proceedings RMRS-P-76, 151-157. Fort Collins, CO: USDA Forest Service. https://www.fs.usda.gov/treesearch/pubs/56726. Accessed 02 February 2022.

Schoettle AW, KS Burns, CM Cleaver, and JJ Connor. 2019a. Proactive limber pine conservation strategy for the Greater Rocky Mountain National Park Area. USDA Forest Service General Technical Report. RMRS-GTR-379. Fort Collins, CO: U.S. Department of Agriculture, Forest Service, Rocky Mountain Research Station. 81 p. https://www.fs.usda.gov/treesearch/pubs/57621. Accessed 02 February 2022.

Schoettle AW, KS Burns, WR Jacobi, J Popp, S Alberts, T Douville, and F Romero. 2018. Southern Rockies Rust Resistance Trial. In Proceedings of the IUFRO joint conference: Genetics of five-needle pines, rusts of forest trees, and Strobusphere, 15-20 June 2014, Proceedings RMRS-P-76, 158-161. Fort Collins, CO: USDA Forest Service. https://www.fs.usda.gov/treesearch/pubs/56727. Accessed 02 February 2022.

Schoettle AW, WR Jacobi, KM Waring, and KS Burns. 2019b. Regeneration for resilience framework to support regeneration decisions for species with populations at risk of extirpation by white pine blister rust. New Forests 50:89-114. https://www.fs.usda.gov/treesearch/pubs/57177. Accessed 02 February 2022.

Schoettle, AW, JG Klutsch, and RA Sniezko. 2012. Integrating regeneration, genetic resistance, and timing of intervention for the long-term sustainability of ecosystems challenged by non-native pests: A novel proactive approach. In Proceedings of the fourth international workshop on the genetics of host-parasite interactions in forestry: disease and insect resistance in forest trees. USDA Forest Service General Technical Report PSW-GTR-240, 112-123. http://www.fs.fed.us/psw/publications/documents/psw_gtr240/psw_gtr240_112.pdf. Accessed 02 February 2022.

Schoettle AW, and RA Sniezko. 2007. Proactive intervention to sustain high-elevation pine ecosystems threatened by white pine blister rust. Journal of Forest Research 12:327-336. https://www.fs.usda.gov/treesearch/pubs/29500. Accessed 02 February 2022.

Schoettle AW, RA Sniezko, A Kegley, and KS Burns. 2011. Preliminary overview of the first extensive rust resistance screening tests of Pinus flexilis and Pinus aristata. In Proceedings of the future of high-elevation, five-needle white pines in Western North America symposium, 28-30 June 2010, Missoula, MT, Proceedings RMRS-P-63, 265-269. USDA Forest Service. https://www.fs.usda.gov/treesearch/pubs/38235. Accessed 02 February 2022.

Schoettle AW, RA Sniezko, A Kegley, and KS Burns. 2014. White pine blister rust resistance in limber pine: Evidence for a major gene. Phytopathology 104:163-173. https://www.fs.usda.gov/treesearch/pubs/44228. Accessed 02 February 2022.

Sniezko RA, R Danchok, DP Savin, JJ Liu, and A Kegley. 2016. Genetic resistance to white pine blister rust in limber pine (Pinus flexilis): major gene resistance in a northern population. Canadian Journal of Forest Research 46(9):1173-1178. doi.org/10.1139/cjfr-2016-0128.

Stephan BR. 2004. Studies of genetic variation with five-needle pines in Germany. In Proceedings of the breeding and genetic resources of five-needle pines: growth, adaptability, and pest resistance conference, 23-27 June 2001, Medford, OR, Proceedings RMRS-P-32, 98-102. USDA Forest Service. https://www.fs.fed.us/rm/pubs/rmrs_p032.pdf. Accessed 02 February 2022.

Vogler DR, A Delfino-Mix, and AW Schoettle. 2006. White pine blister rust in high-elevation white pines: Screening for simply inherited hypersensitive resistance. In Proceedings of the 53rd Western International Forest Disease Work Conference, 26-30 September 2005, Jackson, WY, 73-82. USDA Forest Service Intermountain Region. https://www.fs.usda.gov/treesearch/pubs/29494. Accessed 02 February 2022.

Field Spectroradiometer Methods for Remotely Identifying Treeline Vegetation Species: A Case Study Using Limber Pine in Rocky Mountain National Park

Laurel A. Sindewald, <Laurel.Sindewald@ucdenver.edu>, Department of Integrative Biology, University of Colorado Denver; **Matthew D. Cross**, Department of Environmental Science and Geography, University of Colorado Denver; **Ted Scambos**, Cooperative Institute for Research in Environmental Sciences, University of Colorado Denver; **Diana F. Tomback**, Department of Integrative Biology, University of Colorado Denver; **Peter J. Anthamatten**, Department of Environmental Science and Geography, University of Colorado Denver.

ABSTRACT

Limber pine (*Pinus flexilis)* is an ecologically important conifer in Rocky Mountain National Park (RMNP) that ranges from lower to upper treeline and faces the combined threats of mountain pine beetle (*Dendroctonus ponderosae*) outbreaks, advancing white pine blister rust (*Cronartium ribicola*), and a changing fire regime that could further fragment its metapopulation. As climate warms, limber pine is projected to move upward in elevation into tundra communities. However, the current limber pine distribution at treeline has not been mapped. Here we report groundtruthing methods toward a remote sensing application that might identify the composition of treeline communities by species. First, we used field spectroradiometer methods to characterize intra- and inter-species variation in the spectral reflectance curves of treeline vegetation species. We used photo documentation to connect variation observed in the spectral response to the physical condition of individual trees. We used this field spectroradiometer data to groundtruth WorldView-3 reflectance data toward a future classification of treeline RMNP vegetation.

INTRODUCTION

As a drought-tolerant conifer growing at treeline in Rocky Mountain National Park (RMNP) (Lepper 1980; Steele 1990), limber pine (*Pinus flexilis*) provides a good case study for examining treeline species that may advance under changing climate. Bioclimatic envelope models of limber pine in RMNP predict that its elevational range will move upslope with increasing temperature and precipitation (Monahan et al. 2013). Newer limber pine recruits are infilling treeline communities in RMNP (Sindewald et al. 2020) and limber pine has been observed colonizing at elevations above bristlecone pine (*Pinus longaeva*) in the White Mountains of eastern California (Millar et al. 2018; Smithers et al. 2018). With warming climate, limber pine may tolerate increasing moisture stress and persist on the drier slopes at treeline east of the Continental Divide in RMNP. However, changes would be difficult to distinguish, because the distribution of limber pine at treeline throughout RMNP is not mapped.

Limber pine's distribution comprises a metapopulation structure throughout the Rocky Mountain region (Peet 1978; Webster and Johnson 2000; Williams 2017). Due to this patchy distribution, limber pine may be vulnerable to local extirpation in the event of a large disturbance, such as the Cameron Peak and East Troublesome fires in RMNP in 2020. Though the East Troublesome fire burned up to treeline and jumped the Continental Divide, such treeline fire events are still rare. Treeline stands of limber pine may also escape mountain pine beetles (*Dendroctonus ponderosae*), which prefer larger-diameter stems (Maher et al. 2021), and white pine blister rust, which requires high relative humidity for transmission between alternate hosts and pines (Geils et al. 2010; Thoma et al. 2019). Given enough viable seed production, treeline limber pine may allow the species to persist despite these multiple threats. A systematic inventory of where limber pine is found within the treeline ecotone will allow us to determine which non-climatic, abiotic variables are associated with its occurrence, refining distribution modeling for the species under changing climate. Because limber pine grows on steep slopes and sheer cliffs with difficult access, remote sensing provides a tool to improve mapping and identification.

Aerial photography and satellite imagery have long been used to map treeline position across regions and have helped identify the variables that control treeline (Allen and Walsh 1996; Brown et al. 1994; Leonelli et al. 2016; Wei et al. 2020), but individual tree identification was out of reach until sensor technologies improved, providing higher spatial, radiometric, and spectral resolution. By combining airborne hyperspectral imagery and lidar, tree species can be more easily discriminated (Dalponte et al. 2012; Liu et al. 2017; Matsuki et al. 2015; Shen and Cao 2017; Voss and Sugumaran 2008), in some cases with classification accuracies ranging between 76.5% to 93.2% (Dalponte et al. 2012). Nevertheless, these sensors and aircraft overflights are costly. Satellite remote sensing is less costly, but often comes at the cost of decreased spatial resolution. For example, hyperspectral data from space requires a larger spatial resolution (e.g., the Hyperion satellite, with 220 spectral bands but 30 m spatial resolution), because finer spatial resolution sensors have a smaller instantaneous field of view that collects less radiation, so fewer bands can be resolved. Airborne multispectral imagery also yields high classification accuracy (85.8%) (Dalponte et al. 2012), suggesting that a general threshold of spatial and radiometric resolution can be attained that supports species-level tree identification and mapping.

Recently, a high accuracy discrimination between different species of rainforest trees over a managed forest area within the La Selva Research Center in Costa Rica was accomplished using WorldView-3 (WV-3) imagery (Cross et al. 2019a; Cross et al. 2019b). A field spectroradiometer was used to determine the foliage spectral reflectance curves (light reflectance) of individual tree species. The curves were then compared directly with the spectral reflectance curves observed in the WV-3 imagery after atmospheric correction (Cross et al. 2019a). Two spectral vegetation indices specific to WV-3 bands were developed (Cross et al. 2019b) and used to perform an object-based classification on pixel clusters generated by an image segmentation process. This application achieved high classification accuracy (overall accuracies of 75.61% to 85.37%) (Cross et al. 2019a). Prior to this work, applications of WV-3 imagery for species identification had mixed success and the imagery was often used in combination with machine learning or airborne lidar (Immitzer et al. 2012; Li et al. 2015; Majid et al. 2016; Rahman et al. 2018; Wang et al. 2016).

Our ultimate goal is to use the approach demonstrated by Cross et al. (2019a) to identify where limber pine is located across treeline in RMNP, as an input for future species distribution modeling work. The first step is to determine if we can distinguish limber pine among intermixed tree and shrub species in treeline communities across a highly heterogeneous high-elevation landscape. The application to a treeline ecotone introduces new challenges. In contrast to the La Selva Research Station, which is located on a relatively flat area, treeline terrain in RMNP is extremely rugged, requiring substantial corrections for topography. Conversely, imagery of high-elevation areas with thin atmosphere will require little correction for atmospheric effects such as for the high levels of water vapor in a tropical rainforest. Conifers can be reliably distinguished from broadleaf trees, as was recently demonstrated in low-elevation forests using WV-3 and lidar with 94% accuracy (Varin et al. 2020). Classification of conifer species using WV-3 without the expensive addition of lidar overflight data has not been attempted.

Here we present our methods to obtain spectral profiles of six common treeline species. These methods are prerequisites for using WV-3 imagery to ascertain treeline community composition with a focus on determining the presence of limber pine. We adapted Cross et al.'s (2019a) method to obtain *in situ* spectroradiometer data to estimate interspecific variation of six common treeline vegetation species in RMNP and to groundtruth WV-3 reflectance data. We also added additional *in situ* photographic data to support intra-species variation evaluation, acquiring spectra and corresponding photos of different components of a typical treeline tree (e.g., bare sticks, healthy foliage, dry needles). We then compared spectral curves of whole, geolocated tree canopies obtained with the spectrometer to the reflectance curves extracted from corresponding pixels in the WV-3 satellite imagery to groundtruth the selected pixels.

METHODS

WorldView-3 panchromatic and multispectral imagery was purchased in July 2020 from Maxar covering two study areas in Rocky Mountain National Park where limber pine is abundant at treeline: Longs Peak and Battle Mountain. The Longs Peak study area also includes dense willow (*Salix glauca*, *Salix brachycarpa*, and hybrids), Engelmann spruce (*Picea engelmannii*), subalpine fir (*Abies lasiocarpa*), glandular birch (*Betula glandulosa*), and aspen (*Populus tremuloides*). The Battle Mountain study area is mostly composed of limber pine with Engelmann spruce, subalpine fir, willow, and glandular birch as minor components. The WorldView-3 imagery was collected on July 21, 2020 with 0% cloud cover and an off-nadir angle of 16.8 degrees.

We collected ground control points at trail junctions and switchbacks using a Trimble Geo7x centimeter edition geolocator and a Zephyr 2 antennae on a 1 m carbon fiber pole. These ground control points are select landscape features visible in the satellite imagery. MAXAR used the ground control points to perform orthorectification. Atmospheric correction was done using the MAXAR AComp method, which uses cloud, aerosol, water vapor, ice, and snow (CAVIS) band data collected at the same instance as the multispectral data to identify and correct for specific atmospheric influences (Pacifici 2016).

In July and August 2021, we used an Analytical Spectral Devices (ASD) FieldSpec 4 High Resolution Portable Spectroradiometer to collect spectral reflectance data from each dominant vegetation species present at the two sites. We obtained at least 4-5 replicates of each species. We initially collected data at the Longs Peak site on July 2, August 7, and August 12, and at the Battle Mountain site on August 8. However, on these occasions we did point collections with a 13 cm to 16 cm diameter field of view, which resulted in high variability between collections and high intraspecific variability due to differences in proportions of dead needles and stems. So, we adjusted our methodology and collected at the Battle Mountain site again on September 6. In this last collection, we were limited to sampling one individual of each species.

We collected samples within the four-hour period for two hours on either side of local solar noon (approximately 11 am to 3 pm) to ensure adequate illumination. We were restricted in July to a 1-hour collection period before noon due to thunderstorms. Prior to collections, we turned the spectrometer on for 30 minutes to warm up to stabilize temperature and improve sensor accuracy. We calibrated the spectrometer using a Spectralon white reference placed on the scene.

We ensured that the fiberoptic cable was as close to on-nadir as possible to compare with the satellite perspective, and to collect side-scatter while avoiding fore- or back-scatter. In mid-September, we mounted the fiberoptic cable housing to a hiking pole to hold the cable 1.0 m to 1.5 m above the specimen, on-nadir, for a 0.3 m to 0.5 m diameter field of view. For each individual tree, we collected two replicate samples of green needles, brown stems, and two canopy averages acquired by moving the foreoptic field of view slowly across the tree crown area. The moving canopy averages were carefully planned to maintain an on-nadir view and side-scatter collection as the operator carried the fiberoptic cable over the target tree.

We compared the canopy averages to the WorldView-3 reflectance data. To do this, we converted the spectrometer data (hyperspectral, with reflectance values for every nm wavelength between 250 nm to 2500 nm) to equivalent WV-3 bands. We multiplied each reflectance value by the proportion of total radiation the WV-3 multispectral sensor element can detect for that wavelength, then summed the reflectance values across the WV-3 band range.

We used a Trimble GeoXT and a Trimble Geo7x to record perimeters of individual trees corresponding to the September 6th collections. We imported the data to ENVI (version 4.8, Exelis Visual Information Solutions, Boulder, CO) and used the polygons to select pixels that fell entirely within the bounds of the polygons. We exported that data as csv files and compared the median values for each species to the spectrometer canopy averages with a bootstrapped 95% confidence interval (R, version 4.0.5).

RESULTS

The first collections, on July 2, August 7 to 8, and August 12, were highly variable, particularly in the two near infrared bands, and confidence intervals for the median reflectance for each species overlapped across all WV-3-equivalent bands. This was likely due to the small field of view (13 cm to 16 cm), resulting in large differences introduced by variation in plant condition, with differing proportions of dead needles and twigs or pollen cones.

Our collections on September 6, showed that some of the variability can be replicated by comparing green needle collections to brown needle and moving canopy average collections (figures 1-3). The green samples, aside from showing the expected higher reflectance in the green, also

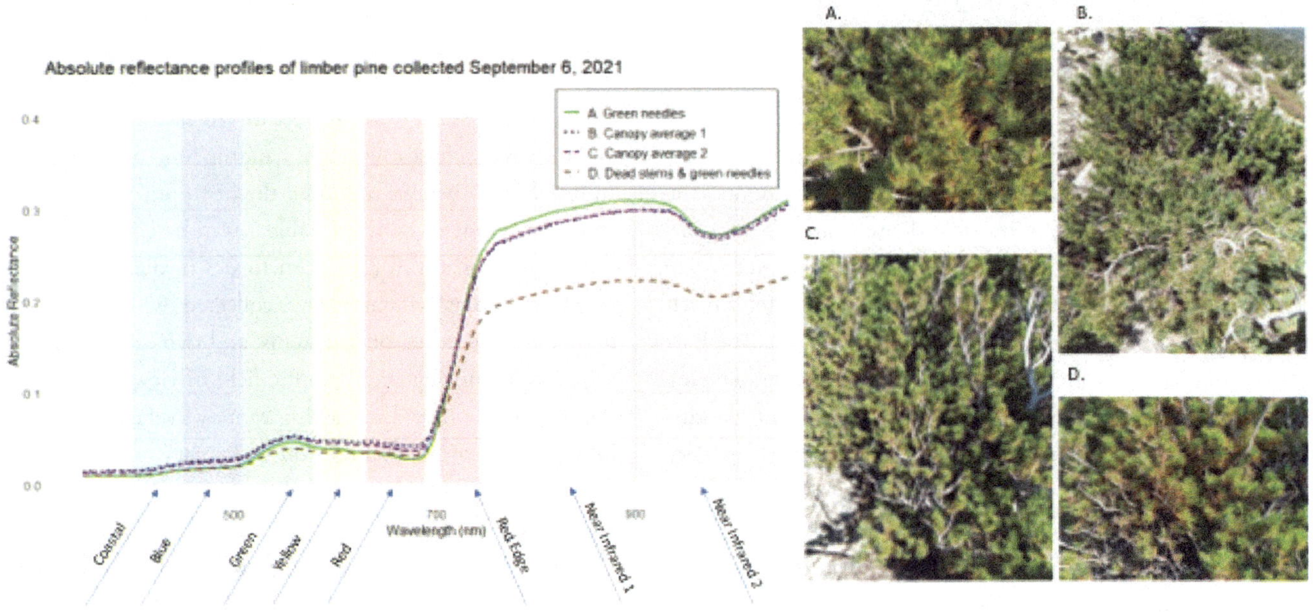

Figure 1. Absolute reflectance profiles of different areas of a krummholz limber pine tree at the Battle Mountain site, collected on September 6, 2021, with an ASD Field Spectroradiometer. Point collections were obtained for green needles and for dead stems and green needles. Moving canopy averages were collected to represent overall variability. The color bands correspond to the wavelength ranges for each of the eight multispectral WV-3 bands. Each labeled collection line in the legend corresponds to a photograph of the sample in the field (A.-D.)

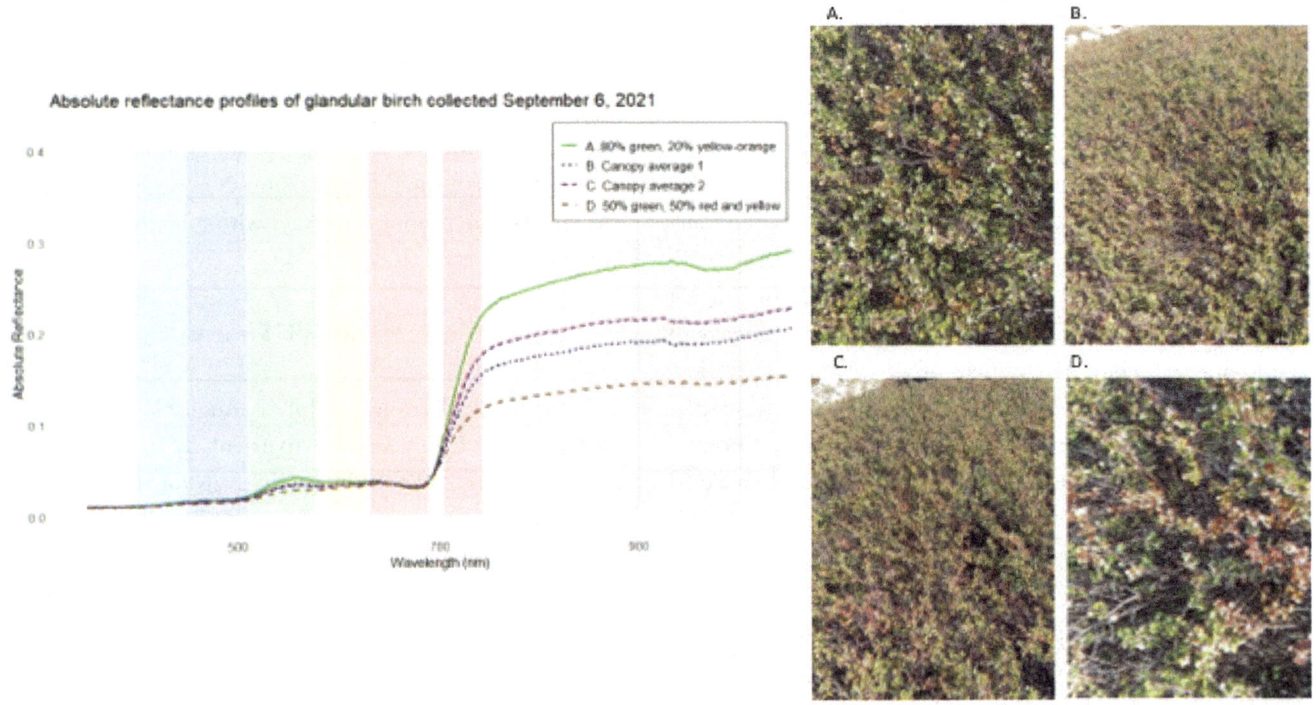

Figure 2. Absolute reflectance profiles of different areas of glandular birch at the Battle Mountain site, collected on September 6. The color bands correspond to the wavelength ranges for each of the eight multispectral WV-3 bands. Each labeled collection line in the legend corresponds to a photograph of the sample in the field.

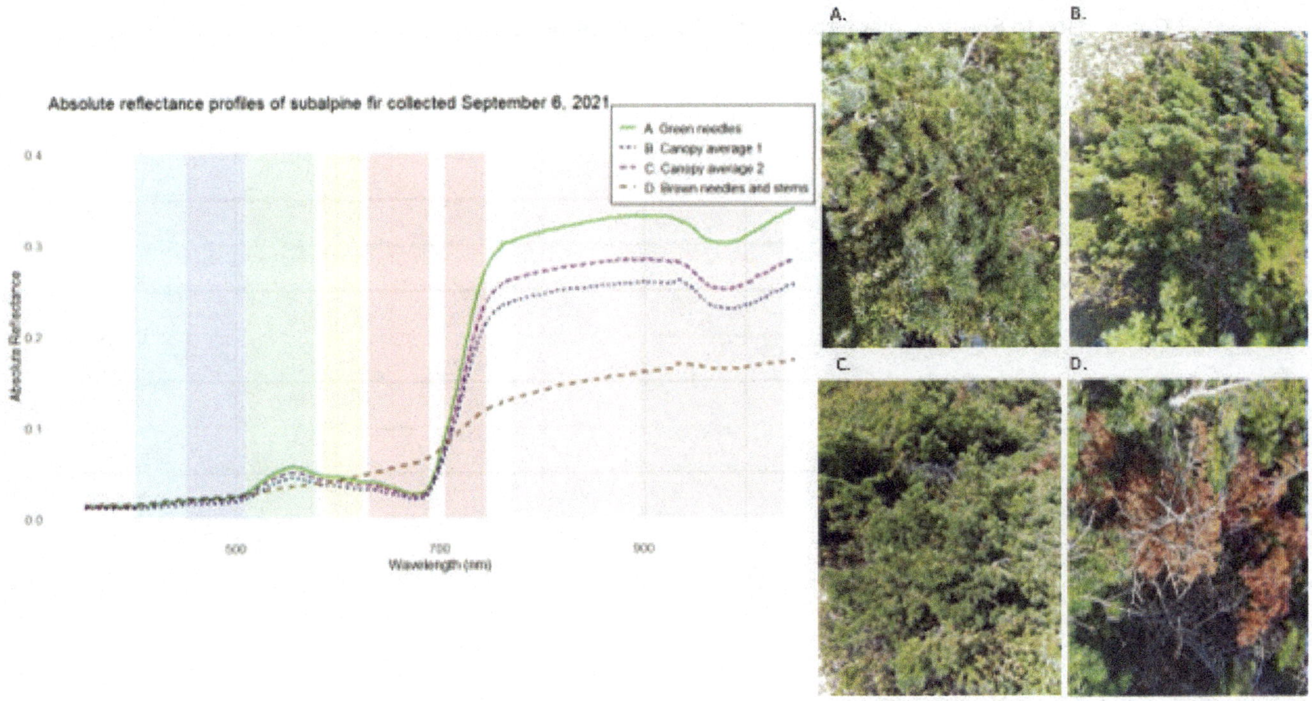

Figure 3. Absolute reflectance profiles of different areas of a krummholz subalpine fir tree at the Battle Mountain site, collected on September 6, 2021. Point collections were obtained for green needles and for dead stems and brown needles. Moving canopy averages were collected to represent overall variability. The color bands correspond to the wavelength ranges for each of the eight multispectral WV-3 bands. Each labeled collection line in the legend corresponds to a photograph of the sample in the field.

showed greater reflectance in both near infrared bands than the samples of brown needles and dead stems. This higher reflectance in the near infrared bands is likely due to water content.

For limber pine, the moving canopy averages were close to the reflectivity values for green needles (figure 1). For glandular birch and subalpine fir, the averages were in between green leaves and dead branches and leaves (figures 2 and 3). By early September, the glandular birch leaves had begun to turn orange and red. Point collections were obtained for an area of mostly green leaves and for an area of mostly red-orange leaves and branches.

The median reflectance values, calculated from the WorldView-3 pixels that fell within the perimeters of the trees or vegetation patches sampled for the spectrometer collections, compared well with the spectrometer reflectance values. The relative difference in reflectance between the bands was mostly consistent between the WV-3 data and the spectrometer data; the shapes of the curves were very close (figures 4 and 5). The absolute reflectance values for each band, however, was greater for the WV-3 data than the spectrometer data.

DISCUSSION

The spectrometer data was successful at groundtruthing the WV-3 imagery, allowing us to compare the training sample data (the reflectance profiles of the pixels in the image for each tree or shrub) to reflectance data collected *in situ*. The spectrometer data also allowed us to demonstrate how each species varied with individual plant condition, useful for interpreting intraspecific variation in reflectance profiles in the satellite imagery. As we attempt a high-resolution satellite imagery classification of the alpine treeline ecotone, field spectrometer data will be invaluable for identifying bands and spectral vegetation indices to discriminate species.

Treeline environments are heterogeneous, with high variability in snow distribution and therefore moisture availability. As plants absorb water, the cytoplasm inside the cell membranes exerts turgor pressure on the cell walls, resulting in structural differences between the cells of plants with greater water content and those of plants that are water stressed. Differences in spectral response due to cell struc-

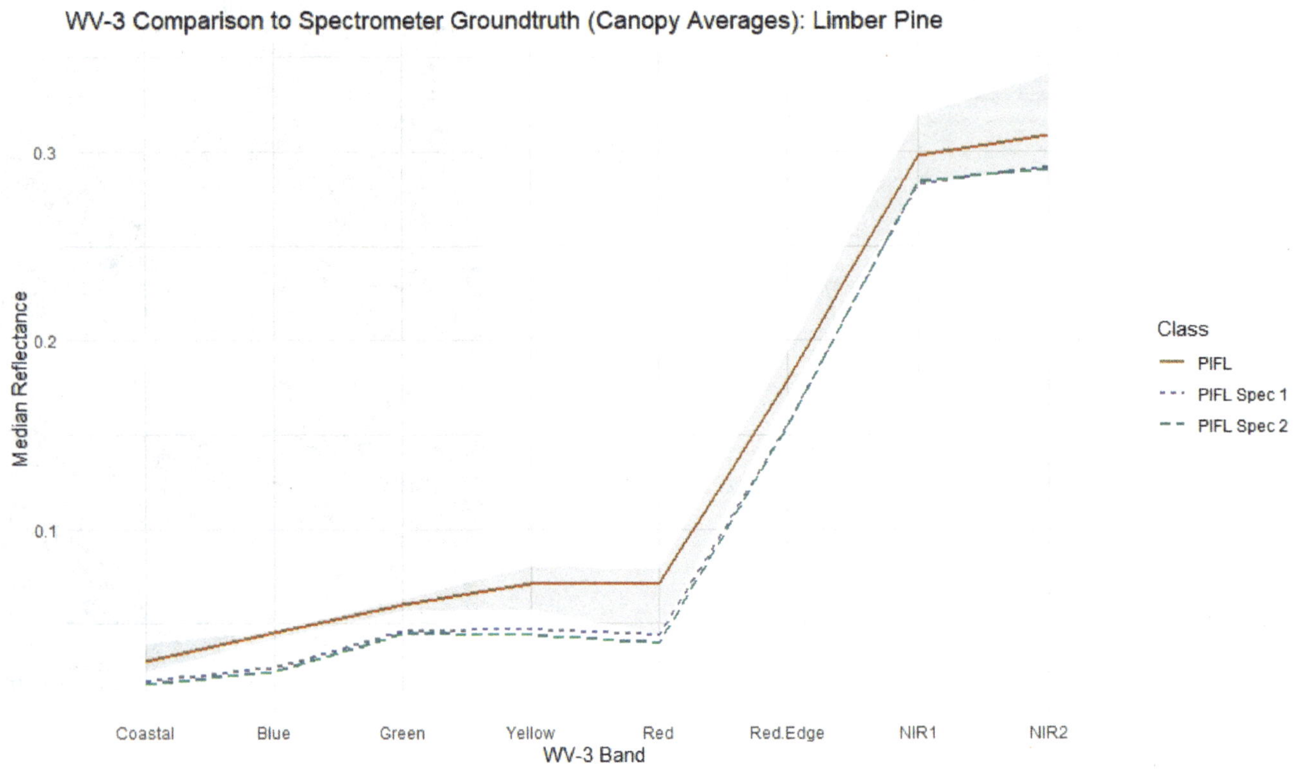

Figure 4. WorldView-3 median reflectance data from pixels that fell within the perimeters of the limber pine (PIFL) tree from which the spectrometer sample was obtained on September 6. The two spectrometer lines are the two canopy averages from the same individual tree, representing overall canopy variation.

ture can be observed in the near infrared region of the electromagnetic spectrum. Therefore, it is likely that high intra- and interspecific variability will be present in the absolute reflectance response of plants at treeline in the near infrared 1 and near infrared 2 (NIR1 and NIR2) WV-3 bands.

The differences observed in NIR1 and NIR2 bands may therefore not be useful for classifying treeline vegetation species, though a larger sample size is necessary to test this. Species may be separable only in the green and red edge bands, though more spectrometer data using the canopy average method is necessary to estimate intraspecies variation.

Individual trees at treeline are also frequently wind and frost-damaged, leading to high levels of intracanopy variation in spectral reflectance. Much of the variation we observed in our first collections, in July and August, is likely attributable to point collections with a small field of view, capturing only a small portion of the overall canopy variability. Point collections are sensitive to differences in the proportion of green needles and brown needles or stems. Unfortunately, this means we cannot use the July and August collections.

As the field of view is moved over the canopy, however, the spectrometer will report an average reflectance for the area covered and will thus represent an average of overall intracanopy variation. The moving canopy average will also more closely represent what the WV-3 sensors pick up at 1.24 m resolution. The shape of the median reflectance curve from the WV-3 imagery for a limber pine tree closely resembles the curves obtained for the two canopy averages (figure 4). A departure in curve shape for glandular birch (figure 5) is seen in the red edge and near infrared bands, likely due to deciduous leaf phenology. As the leaves change color and die, the cells collapse and the leaves fall, resulting in a greater proportion of stems and dead leaves in both canopy averages and lower reflectivity in the near infrared.

Collecting moving averages for canopies could allow for the development of spectral libraries for species, representing the variability of species reflectance across the electromagnetic spectrum. A greater representation of in-

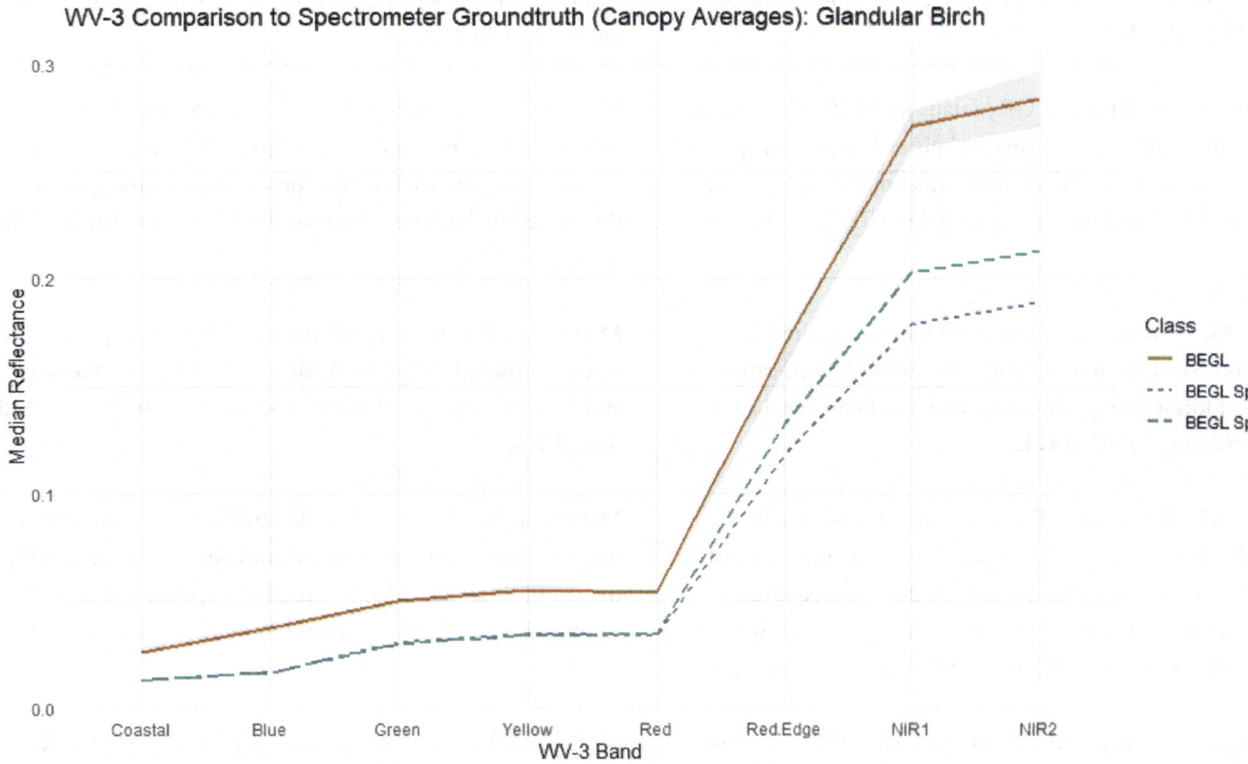

Figure 5. WorldView-3 median reflectance data from pixels that fell within the perimeters of the glandular birch (BEGL) patch from which the spectrometer sample was obtained on September 6. The two spectrometer lines are the two canopy averages from the same patch, representing overall variation.

tra-species variation could then lead to statistical rules for classification of each species on new imagery.

MANAGEMENT IMPLICATIONS

Our paper provides a detailed description of methodology for obtaining an *in-situ* spectral groundtruth of treeline vegetation for imagery classification applications. Spectrometer data provides a spectral groundtruth for training samples identified in the imagery, enabling researchers to determine whether the training pixels identified in the imagery are correctly aligned with objects on the ground. Spectroradiometer groundtruthing may also improve our understanding of the sources of variability that lead to poor classification accuracy, such as variable access to water resources at treeline. Finally, the development of a spectral library of vegetation species, fully representing intraspecies variation over time, may lead to statistical rule sets that could then be used to distinguish species in new locations in new imagery, much as one can use satellite imagery to identify the probable locations of large mineral deposits of specific types. The rise in availability of affordable, high-resolution satellite imagery may then lead to more efficient monitoring of plant species migration and land cover changes under novel climate scenarios.

ACKNOWLEDGEMENTS

The Office of Research Services, University of Colorado, Denver, provided funding to DFT, MDC, and PJA for this research. The Continental Divide Research Learning Center in Rocky Mountain National Park provided lodging for our researchers. We would also like to thank our field assistants—Lucas Rudisill, Ryan Lagerquist, and Kyla Lucketta—for their work at treeline.

LITERATURE CITED

Allen TR, and SJ Walsh. 1996. Spatial and compositional pattern of alpine treeline, glacier national park, montana. Photogrammetric engineering and remote sensing 62(11):1261-1268.

Brown DG, DM Cairns, GP Malanson, SJ Walsh, and DR Butler. 1994. Remote sensing and gis techniques for spatial and biophysical analyses of alpine treeline through process and empirical models. United Kingdom: Taylor & Francis Ltd.

Cross M, T Scambos, F Pacifici, O Vargas-Ramirez, R Moreno-Sanchez, and W Marshall. 2019a. Classification of tropical forest tree species using meter-scale image data. Remote Sensing 11(12):1411.

Cross MD, T Scambos, F Pacifici, and WE Marshall. 2019b. Determining effective meter-scale image data and spectral vegetation indices for tropical forest tree species differentiation. IEEE Journal of Selected Topics in Applied Earth Observations and Remote Sensing 12(8):2934-2943.

Dalponte M, L Bruzzone, and D Gianelle. 2012. Tree species classification in the southern alps based on the fusion of very high geometrical resolution multispectral/hyperspectral images and lidar data. Remote Sensing of Environment 123:258-270.

Geils BW, KE Hummer, and RS Hunt. 2010. White pines, ribes, and blister rust: A review and synthesis. Forest Pathol 40(3-4):147-185.

Immitzer M, C Atzberger, and T Koukal. 2012. Tree species classification with random forest using very high spatial resolution 8-band worldview-2 satellite data. Remote Sensing 4(9):2661-2693.

Leonelli G, A Masseroli, and M Pelfini. 2016. The influence of topographic variables on treeline trees under different environmental conditions. Physical Geography 37(1):56-72.

Lepper MG. 1980. Carbon dioxide exchange in pinus flexilis and p. Strobiformis (pinaceae). Madroño 27(1):17-24.

Li D, Y Ke, H Gong, and X Li. 2015. Object-based urban tree species classification using bi-temporal worldview-2 and worldview-3 images. Remote Sensing 7(12):16917-16937.

Liu L, NC Coops, NW Aven, and Y Pang. 2017. Mapping urban tree species using integrated airborne hyperspectral and lidar remote sensing data. Remote Sensing of Environment 200:170-182.

Maher CT, CI Millar, DLR Affleck, RE Keane, A Sala, C Tobalske, AJ Larson, and CR Nelson. 2021. Alpine treeline ecotones are potential refugia for a montane pine species threatened by bark beetle outbreaks. Ecological Applications n/a(n/a):e2274.

Majid IA, ZA Latif, and NA Adnan. 2017. Tree species classification using worldview-3 data. 2016 7th IEEE Control and System Graduate Research Colloquium (ICSGRC); 8-8 Aug. 2016.

Matsuki T, N Yokoya, A Iwasaki. 2015. Hyperspectral tree species classification of Japanese complex mixed forest with the aid of lidar data. IEEE Journal of Selected Topics in Applied Earth Observations and Remote Sensing 8(5):2177-2187.

Millar CI, DA Charlet, DL Delany, JC King, and RD Westfall. 2019. Shifts of demography and growth in limber pine forests of the Great Basin, USA, across 4000 yr of climate variability. Quaternary Research 91(2): 691-704.

Monahan WB, T Cook, F Melton, J Connor, and B Bobowski. 2013. Forecasting distributional responses of limber pine to climate change at management-relevant scales in Rocky Mountain National Park. Plos One. 8(12).

Pacifici F. 2016 Validation of the digitalglobe surface reflectance product. 2016 IEEE International Geoscience and Remote Sensing Symposium (IGARSS); 10-15 July 2016.

Peet RK. 1978. Latitudinal variation in southern Rocky Mountain forests. Journal of Biogeography 5(3):275-289.

Rahman MM, A Robson, and M Bristow. 2018. Exploring the potential of high resolution worldview-3 imagery for estimating yield of mango. Remote Sensing 10(12):1866.

Shen X, and L Cao. 2017. Tree-species classification in subtropical forests using airborne hyperspectral and lidar data. Remote Sensing 9(11):1180.

Sindewald LA, DF Tomback, ER Neumeyer. 2020. Community structure and functional role of limber pine (pinus flexilis) in treeline communities in Rocky Mountain National Park. Forests 11(8):838.

Smithers BV, MP North, CI Millar, and AM Latimer. 2018. Leap frog in slow motion: Divergent responses of tree species and life stages to climatic warming in Great Basin subalpine forests. Global Change Biology 24(2):E442-E457.

Steele R. 1990. Pinus flexilis James. Silvics of North America. Washington, DC: US Department of Agriculture. p. 348-354.

Thoma DP, EK Shanahan, and KM Irvine. 2019. Climatic correlates of white pine blister rust infection in whitebark pine in the Greater Yellowstone Ecosystem. Forests 10(8):666.

Varin M, B Chalghaf, and G Joanisse. 2020. Object-based approach using very high spatial resolution 16-band worldview-3 and lidar data for tree species classification in a broadleaf forest in Quebec, Canada. Remote Sensing 12(18):3092.

Voss M, and R Sugumaran. 2008. Seasonal effect on tree species classification in an urban environment using hyperspectral data, lidar, and an object-oriented approach. Sensors 8(5):3020-3036.

Wang T, H Zhang, H Lin, and C Fang. 2016. Textural–spectral feature-based species classification of mangroves in Mai Po Nature Reserve from worldview-3 imagery. Remote Sensing 8(1):24.

Webster KL, and EA Johnson. 2000. The importance of regional dynamics in local populations of limber pine (pinus flexilis). Ecoscience 7(2):175-182.

Wei C, DN Karger, and AM Wilson. 2020. Spatial detection of alpine treeline ecotones in the western United States. Remote Sensing of Environment 240:111672.

Williams TJ. 2017. Clark's nutcracker seed use and limber pine metapopulation structure in Rocky Mountain National Park. [Denver, CO]: University of Colorado Denver.

Repeat Photography Helps Tell the Story of Whitebark Pine

Jane Kapler Smith, <mtjanek@yahoo.com>, USDA Forest Service, Rocky Mountain Research Station, Missoula Fire Sciences Laboratory (retired); **Garon C. Smith**, Department of Chemistry (emeritus), University of Montana; Stephen F. Arno, USDA Forest Service, Rocky Mountain Research Station, Missoula Fire Sciences Laboratory (retired).

ABSTRACT

Repeat photography—that is, photos taken from exactly the same location over time—is a powerful tool for describing landscape changes over time. However, this technique has not been used widely to document the decline of five-needle pines nor for showing progress in their restoration. We used repeat photography to show mortality of whitebark pine (Pinus albicaulis) at several locations in western Montana, USA; three of the resulting photo pairs are included in this paper. Because whitebark pine has declined so rapidly, we did not need the original, historical images to be very old. Even photos taken 16 years apart show dramatic changes in whitebark pine forests. Our photo pairs illustrate the heart-breaking story of whitebark pine mortality. More hopeful ways to use repeat photography would be to choose sites for restoration of whitebark pine that have a historical photographic record of healthy trees, and to establish new photo points on sites where whitebark pine restoration is underway.

INTRODUCTION

A pair or series of pictures taken from one location over time ("repeat photography") can be worth a thousand words for describing ecological change. We used this technique to illustrate the widespread loss of whitebark pine (*Pinus albicaulis*) during the past 50 years. While the whitebark pine's decline is not news (see, for example, Arno et al. 1993; Keane and Arno 1993), repeat photography has not been used widely to tell this heart-breaking story. The technique can be valuable in many ways: It can vividly illustrate the extent and rapidity of whitebark decline. It can provide context for sampling site data, since it can show change through time over a large area. Finally, if used in areas where whitebark pine is being restored, it can convey hope for the future of this beautiful species and the hundreds of species in its habitat - especially the iconic Clark's nutcracker (*Nucifraga columbiana*).

Repeat photography has been used to illustrate the effects of succession and disturbance in many ecosystems. Wright and Bunting (1994) used repeat photography to illustrate change in the abundance and health of limber pine (*Pinus flexilis*) at Craters of the Moon National Monument. White and Hart (2007) used hundreds of repeat photos to depict landscape change over time in the northwestern United States and southwestern Canada. Many of their photos are included in White's (no date) "Lens of Time Northwest" website, which uses computer animation—fading from a historical photo to a more recent one, then back again—to dramatize landscape change. The website is now used in a learning activity that challenges students to use photo pairs and dendrochronological data to envision how a given forest may change in the future (Smith et al. 2018). Similarly, Abrahamson et al. (2017) incorporated Gruell's (2001) repeat photographs from the Sierra Nevada into a learning activity for students in northern California.

An ecological restoration project in a managed ponderosa pine/Douglas-fir (*Pinus ponderosa*/*Pseudotsuga menziesii*) forest (Smith and Arno 1999) shows how repeat photography can complement historical reports and plot data. Treatments were conducted in the Lick Creek Demonstration and Research Forest, Bitterroot National Forest, Montana, USA. They were based on Gruell and others' (1982) repeat photographs and synthesis of historical records from the area. The 1999 report's section on "Educational Value" notes that repeat photos provide visible evidence of the abstract concept of succession. Viewers can trace changes in a single location through time, following growth or loss of individual trees and noting the development of a dense understory. Thus, repeat photography makes the concepts of succession clear and the development of ladder fuels vivid. The Lick Creek project is continuing, with a website providing documentation and links to other products (Missoula Fire Sciences Laboratory (no date)).

METHODS

We used repeat photography to show changes in several whitebark pine stands in western Montana, USA. While most repeat photography projects show changes spanning many decades and sometimes more than a century, the decline of whitebark pine has been so rapid—and we are so old—that we could rephotograph some of our own photos from past years. This also made it relatively easy to find the places where the original photos were taken ("photo points"). The most useful images were those with a distinct foreground and background. By lining up foreground objects with background objects (or the horizon), we could orient the repeat photo correctly and pinpoint the original photographer's location with a few meters. The original photos needed to be clear, and it was helpful if they did not contain high contrast caused by patchy cloud cover, since this contrast could obscure any differences between cover types on the ground.

After selecting photos, we studied maps and satellite images to better identify the photo points, printed copies of the photos, then traveled to the sites and hoped for good weather. When we found the photo points, we recorded latitude and longitude and took multiple photos in an effort to capture the same field of view as that shown in the original. Since none of the repeat photos were taken with the same lens as the originals, they did not match perfectly, so we later cropped and adjusted them to minimize differences in the field of view.

In this paper, we use side-by-side pairs of photos to show the changes that have occurred in whitebark pine communities. In presentations and online, we have used computer animation instead. This provides an even more striking way to depict the decline of whitebark pines. We used PowerPoint to create the animations: The original photo fills the whole screen; then the repeat photo fades in on top of the first, usually from right to left, over a 10-second interval. As viewers go back and forth, they can watch the changes unfolding in different parts of the photos.

RESULTS AND DISCUSSION

Sad to say, it does not take a century or more to see dramatic change in whitebark pine ecosystems. Our photo pair with the longest time interval, 49 years (figure 1), shows a cabin near the eastern edge of Martin Lake in the Deer Lodge National Forest, MT, west of Mount Powell. The cabin is built from whitebark pine logs. In 1971, it was surround-

Figure 1. Left (1971): Whitebark pines surrounding cabin at Martin Lake, Deerlodge National Forest, MT (46° 21' 12.3" N, 112° 59' 36.5" W, elevation 2637 m (8651 ft)). Right (2020): After 49 years, most of the whitebark pines had died and many had fallen. Subalpine fir and alpine larch were becoming established. (Left photo: Steve Arno. Right photo: Garon Smith.)

Figure 2. Left (1989): Scotch Bonnet Mountain, northeast of Lulu Pass, Gallatin National Forest, MT (45° 02' 52.5" N, 109° 55' 41.0" W, elevation 2878 m (9442 ft)). Right (2018): After 29 years, patches of dead trees—mostly whitebark pines—were visible on the slopes to the right of Scotch Bonnet Mountain and along the left edge of the photo. Tree regeneration in the foreground included only one whitebark pine. (Left photo: Steve Arno. Right photo: Garon Smith.)

Figure 3. Left (2005): Whitebark pines on ridge northwest of Glen Lake, Bitterroot National Forest, MT (46° 27' 35.7" N, 114° 17' 54.9" W, elevation 2607 m (8554 ft)). Whitebark pines were already declining. Right (2021): After 16 years, nearly all of the whitebarks were dead. (Both photos by Garon Smith.)

ed by large whitebark pines. In 2020, the cabin was still standing—albeit with a lean and no roof—but most of the whitebark pines were dead and many had fallen. Whitebark mortality was high throughout the area, but a few vigorous, cone-bearing trees were present. Subalpine fir (*Abies lasiocarpa*) and alpine larch (*Larix lyallii*) were abundant.

A second photo pair (figure 2) spans a 29-year interval. The photo point is on the northeastern slopes of Henderson Mountain in the Gallatin National Forest, MT, looking northwest at Scotch Bonnet Mountain. Lulu Pass is on the far horizon on the left, and part of Sheep Mountain is shown on the right. In 1989, forest cover was nearly continuous on the slopes below Scotch Bonnet and Sheep Mountains, although dead tree crowns were also visible. In 2018, large patches of dead trees were visible on these slopes, and patches of dead trees were also evident along the left edge of the photo. Substantial tree regeneration was visible in the foreground, but it included only one whitebark pine.

A third photo pair (figure 3) shows dramatic change over just 16 years. The photo point is at the top of the ridge northwest of Glen Lake in the Bitterroot National Forest, MT. Whitebark pines were already declining in 2005, but some of the tree crowns were still intact and many young, cone-bearing whitebark pines were present in the area. In 2021, all of the large whitebark pines were dead, and only a few young pines were present.

All of our photo pairs and our PowerPoint animations will be available for download from the Whitebark Pine Foundation's website (www.whitebarkfound.org). We hope other photographers will contribute photos as well, so future visitors to high-elevation forests can continue to tell the whitebark pine story.

LIMITATIONS AND RECOMMENDATIONS

The original photographs for our project were not necessarily taken for the purpose of repeat photography, so our project has inherent limitations, as does repeat photography in general. First, our photo pairs do not necessarily represent the condition of whitebark pine stands throughout western Montana, so they cannot be used as a basis for generalization. Second, the photo pairs are qualitative rather than quantitative; it is difficult to quantify trends from photos alone, and we did not attempt to do so. No quantitative information was available on the historical condition of the areas photographed, and we did not collect data ourselves. Third, photo points that show a landscape provide little information on conditions at a fine scale, such as understory cover and surface fuels.

Repeat photography could complement quantitative studies in whitebark pine. Photos can capture a much larger area than most sample plots, thus providing a visual context for plot data. Repeat photography is essential for documenting restoration projects and communicating their effectiveness. We can generally grasp a story told in photos more easily and quickly than we can interpret quantitative information that is summarized in graphs and tables. Photo points should be established at the beginning of the project and rephotographed as part of the monitoring protocol. To maximize the value of photo points, each one should be described by latitude, longitude, azimuth, a description of the physical setting, and possibly directions for accessing the site. It may be possible to select some restoration sites based on the availability of old photos that show healthy whitebark pines. In that case, photo pairs that currently show only decline could be turned into trios and quartets that show promise for the future of whitebark pine and the organisms that depend on it.

CONCLUSIONS

Whitebark pines are an essential component of healthy, resilient subalpine ecosystems in this region, but few people visit them, understand their fragility, or appreciate their beauty. Repeat photography can help viewers understand the extent and rapidity of whitebark pine's decline. It can also help those who love and care for whitebark pine to communicate the urgent need for protecting and restoring these landscapes. Repeat photography on sites where whitebark pine is regenerating can tell stories of hope, providing much-needed good news to managers and the public.

More than a century ago, poet Gerard Manley Hopkins expressed his grief at the loss of a row of beautiful trees lining a road in rural England, mourning that "After-comers cannot guess the beauty been" (Hopkins 1879). Repeat photography of whitebark pine stands can help "after-comers" appreciate what has been lost in high-elevation forests during the past century and envision the possible restoration of these unique, invaluable ecosystems.

ACKNOWLEDGMENTS

Thank you to Rebecca and Brendan Collins Jordan, Ned Vasquez, and Breanna McCabe for helping us search for photo locations. Thank you to Helen Smith and Penny Morgan for their thoughtful, very helpful reviews.

LITERATURE CITED

Abrahamson, I, JK Smith, and C Berkowitz. 2017. Chapter 19. Sierra Nevada Forests Today. In: FireWorks curriculum featuring lower and upper Sierra Nevada mixed conifer forests. Missoula, MT: USDA Forest Service, Rocky Mountain Research Station (producer). Available: https://www.frames.gov/documents/fireworks/curriculum/Sierra-Nevada/HighSchool/H19/H19_SierraNevadaForestsToday.pdf. Accessed 29 November 2021.

Arno, SF, ED Reinhardt, and JH Scott. 1993. Forest structure and landscape patterns in the subalpine lodgepole pine type: a procedure for quantifying past and present conditions. General Technical Report INT-294. Ogden, UT: U.S. Department of Agriculture, Forest Service, Intermountain Research Station. 17 p.

Gruell, GE. 2001. Fire in Sierra Nevada forests: A photographic interpretation of ecological change since 1849. Missoula, MT: Mountain Press Publishing Company. 238 p.

Gruell, GE, WC Schmidt, SF Arno, and WJ Reich. 1982. Seventy years of vegetative change in a managed ponderosa pine forest in Western Montana—Implications for resource management. Gen. Tech. Rep. INT-130. Ogden, UT: U.S. Department of Agriculture, Forest Service, Intermountain Forest and Range Experiment Station. 42 p.

Hopkins, GM. 1879. Binsey Poplars. https://www.poetryfoundation.org/poems/44390/binsey-poplars. Accessed 29 November 2021.

Keane, RE, and SF Arno. 1993. Rapid decline of whitebark pine in western Montana: evidence from 20-year remeasurements. Western Journal of Applied Forestry 8(2): 44-47.

Missoula Fire Sciences Laboratory. [No date]. A Century of Change in a Ponderosa Pine Forest. https://www.firelab.org/project/century-change-ponderosa-pine-forest. Accessed 29 November 2021.

Smith, HY, and SF Arno, eds. 1999. Eighty-eight years of change in a managed ponderosa pine forest. General Technical Report RMRS-GTR-23. Ogden, UT: USDA Forest Service, Rocky Mountain Research Station. 55 p. https://www.fs.fed.us/rm/pubs/rmrs_gtr023.pdf. Accessed 29 November 2021.

Smith, JK, I Abrahamson, C Berkowitz, and N McMurray. 2018. Chapter 22. Changing Landscapes, Changing Fires. In: FireWorks curriculum featuring ponderosa, lodgepole, and whitebark pine forests. Missoula, MT: USDA Forest Service, Rocky Mountain Research Station, Missoula Fire Sciences Laboratory (Producer). Available: https://www.frames.gov/documents/fireworks/curriculum/NRockies-NCascades/HighSchool/H22/H22_ChangingLandscapesChangingFires.pdf. Accessed 29 November 2021.

White, C. [No date]. Lens of time northwest: Repeat photography, then and now images. https://lensoftimenorthwest.com. Accessed 29 November 2021.

White, C, and EJ Hart. 2007. The Lens of Time: A Repeat Photography of Landscape Change in the Canadian Rockies. Calgary, AB: University of Calgary Press. 320 p.

Wright, RG, and SC Bunting. 1994. The landscapes of the Craters of the Moon National Monument. Moscow, ID: University of Idaho Press. 103 p.

Whitebark Pine Community Processes, Environment and Human Influences: Revisiting Montana State University Work of 1971-2000

T. Weaver, <tWeaver@montana.edu>, Ecology Department, Montana State University, Bozeman, MT.

ABSTRACT

I revisit 30 years of whitebark pine (WBP, *Pinus albicaulis*) reflecting work reported in >30 WBP papers. Using a sample of 47 stands from the northern Rocky Mountains, I document the remarkable openness of WBP stands. We show strong self-thinning and seedling failure, related to soils, not light, and ongoing production/decomposition though 600 years. WBP's remarkable multiple-stemmed trees arise from seed caches/polyembryony combined with lack of light competition. WBP's usual timberline range seems limited upward by growing season length, wind, growing season length, and soil condition and downward by competition, none much directly temperature related. Human impacts discussed include foraging, trampling/compaction, exotic species invasion, and white pine blister rust, which may be significantly ameliorated by introduction of genes from resistant European stone pine species.

INTRODUCTION

I revisit whitebark pine (WBP, Pinus albicaulis) studies made by my plant ecology lab during the 1971- 2000 period in the northern Rocky Mountains. Objectives are keying the availability of useful information, freshly integrating many observations, and speculating on/pondering unresolved problems. A fourth objective is to document, as requested by the Hi5 historical section, early work in the field. This is done in greater depth by Weaver (2022). This review emphasizes the plant ecology lab's ~34 papers and focuses on the work of my fine associates (D Dale, F Forcella, J Jacobs, J Keck, W McCaughey, J Lichthardt, D Gustafson) and of Wyman Schmidt (USDA Forest Service, Forestry Sciences Laboratory, Bozeman) who was so important to the organization and reporting of early WBP research.

I divide my review into three segments: 1) the WBP forest and its dynamics, 2) the response of WBP to its environment, and 3) human impacts on the WBP forest.

THE WBP FOREST AND ITS DYNAMICS

Forest Structure and Dynamics

A typical WBP forest is an open stand of multi-stemmed five-needled trees. Its structure is broadly described by comparing forty-seven stands ranging from 0-600 years (Weaver et al. 1990). Canopy cover increases rapidly from zero at initiation to saturation (~60%) at 100-200 years and then plateaus (table 1). Leaf area, indexed by the sum of circumferences of all component trees, rises more rapidly and levels off similarly. Simultaneously, canopy height increases steeply to 12 m at 200 years and levels off (Weaver and Dale 1974).

Stand density is initially ~1700 seedlings ha-1, peaks soon at ~4600 seedlings ha^{-1} and declines exponentially to 50 ha^{-1} at 400 years (table 1). The initial rise indicates continued seeding from external sources into forests too young to produce their own seed. The number of resultant small trees (0-10 cm diameter at breast height; dbh) declines from establishment to near 50 ha^{-1} at 400 years. The disappearance of these trees must be due either to death and decomposition in place or to graduation into the larger size class (10-90 cm dbh). The large trees also begin to thin after 200 years.

Basal area ($\sum\pi$ $(dbh/2)^2$, an index of standing crop) rises rapidly over the first 100-200 years, then growth apparently tapers (table 1 and Forcella and Weaver 1977) due to the sum of two processes: loss of dying trees and ongoing diameter growth of living trees. Despite the death or graduation of most small trees, the total basal area of small trees is maintained by ongoing diameter growth, which would be impossible without the plentiful light of the open forest. Similarly, the basal area of large trees increases constantly, despite the death of one-third of them, between 400-600 years. The pattern and magnitudes are very similar to those modeled by Keane et al. (1990). The change in basal area and standing crop over time (production) was measured much more elegantly (Forcella and Weaver 1977). This paper partitioned biomass/growth into leaf, branch, bole, root, and bark components—the latter two rarely measured—and shows that due to their long lives, WBP forests can be more massive than those of temperate forests at lower altitudes or in other areas.

The large number of disappearing trees—1450 ha^{-1} small ones between years 100-400 and 450 ha-1 large ones between years 200-400—calls our attention to the fate of the missing trees, a rare subject in forestry. The understory of WBP forests is trash-free and thus, must harbor a potent decomposing agent. Because regular ground fires are not reported, we speculate that fungal decomposition is rapid under long-lying moist-warm (0°C) snow packs. The depth of snow packs in lower altitude WBP forests may be near 2 m, i.e. as indicated by the height to which lichens are excluded from tree trunks (Eversman et al. 1990), possibly by the relatively warm, moist conditions in the snow packs.

Current reproductive effort was estimated by counting cones in the canopy. To compare production across a series of preceding years, we analyzed branch samples by recording cone production of sample branches at progressively earlier times. This was done by recording, for each branch, numbers of juvenile cones, current cones, and scars left by previous cones at annual nodes from the preceding 5-11 years. Average cone crops were near 1.5 m^{-2} for stands older than 100 years (Weaver and Forcella 1986). High cone yields tended to follow low yields, i.e. the trees exhibited masting behavior. Ecologically, masting partially controls seed-consuming mammal populations, such as sedentary red squirrels (*Tamiasciurus hudsonicus*) and chipmunks (*Tamias* spp.), but has less effect on mobile seed predators, such as Clark's nutcrackers (*Nucifraga columbiana*). We could not identify weather conditions that correlated well with yields, perhaps because too few weather stations were located within WBP stands for meaningful results. Yields were inconsistent region-wide, thus usually providing cross-stand dispersal opportunities.

The understory of the WBP/whortleberry association, is dominated by a low shrub (whortleberry, *Vaccinium scoparium*) and may contain 5-10 species of low-density herbs, including glacier lily (*Erythronium grandiflorum*), elk sedge (*Carex geyeri*) and arnica (*Arnica* spp.) (Weaver and Dale 1974, Forcella 1978). The crustose lichen *Parmeliopsis* occupies the bases of many trees and, above 2 m, are fruticose lichens (*Letharia*, *Usnea*, and *Bryoria*), leafy lichens (*Melenelia*), and crustose lichens (*Parmeliopsis* and *Lecanora*) (Eversman et al. 1990). The macrofungal (mushroom) richness may be near 60 species (Keck 2001).

Table 1. Whitebark pine forest dimensions. Values from Weaver, Forcella and Dale (1990).

	CANOPY			DENSITY (trees/ha)		BASAL AREA (m^2/ha)		
Stand age (years)	Cover%	Σ.Circa[a]	Seedling	0-10 cm dbh	>10 cm dbh	0-10 cm dbh	>10 cm dbh	Total
0	0	0	0	0	0	0	0	0
100	40	70	1000	1500	1000	24	14	38
200	57	75	1000	200	1300	25	32	57
–								
400	60	75	50	50	850	18	40	58

[a]Leaves and branches are connected directly to roots by elements in the sapwood, thus sapwood on the circumference correlates well with leaf area. Sapwood is expressed as the sum of circumferences, m/ha.

Table 2. Comparison of environments of major Rocky Mountain environmental zone habitat types (HT). Relatively high values are shaded.

	Climate[b]		Water holding capacity[c]					Nutrients[d]			
HT[a]	T	AP	WHC	depth	Clay	Sand	Org C	N	Ca	K	Mg
	°C	cm	cm	dm	%	%	T/ha	T/ha	T/ha	T/ha	T/ha
Alpine	8	106	3.8	5	90	10	124	10	3	0.3	0.5
Abla/Pial	10	82	4.4	22	96	4	35	3	2	0.8	1
Psme	12	58	10.3	13	82	18	45	4	15	2	5
Feid	13	38	10.1	5	73	27	130	11	27	1.9	6
Bogr	16	35	11.7	6	70	30	82	9	33	2.7	7

[a]HTs: Alpine, conifer forests (moist *Abies lasiocarpa*, ABLA and drier *Pseudotsuga menziesii*, PSME), grasslands (moist *Festuca idahoensis*, FEID and dry *Bouteloua gracilis*, BOGR)

[b]Climate: growing season temperature, T (Weaver 1980), annual precipitation, AP (Weaver 1980).

[c]Water holding capacity (WHC): A&B horizons, and the determinants, soil depth, clay, sand, and organic carbon.

[d]Nutrients: metric tons per hectare.

Multiple Stems and the Stand Life Cycle

WBP forest is characterized by its apparently multi-stemmed trees, as well as by the openness of the canopy discussed above. Two forces are involved. First, multi-stems are actually separate individuals arising from seeds planted in groups (1-3 most commonly but as many as 14) by seed-caching Clark's nutcracker (Hutchins 1990). Regardless of the number of seeds in a cache, clump sizes may be increased by the occurrence of multiple seedlings/stems arising from single seeds. Single seeds commonly produce multiple seedlings (one to >5) but most commonly three (Weaver and Jacobs 1990). Second, and equally important, the openness of the forest eliminates the competition that prevents self-thinning following close establishment of trees arising from caches or multiple embryos. As a confirming aside, I note that other tree species, e.g. Douglas fir (*Pseudotsuga menziesii*) and lodgepole pine (*Pinus contorta*), also may exhibit multiple stems in open stands that lack within-canopy competition.

As a result of both the rarity of seedling establishment and the planting of trees by nutcrackers, I propose, in four statements, that WBP forests are 'climax/self-sustaining' in a somewhat unusual sense, i.e. perhaps only when associated with Clark's nutcracker. First, observation of an altitudinal transect (Weaver 1990b) shows a clean high forest with no invasion by subalpine fir; i.e. it seems to be a climax species. This is in contrast to seral lodgepole pine forests below it, which are regularly replaced by subalpine fir (cf. Pfister et al. 1977). Second, the forest dynamics section above shows that the forest evolves in an orderly way through 600 years without replacement by other tree species. Third, while WBP produces abundant seeds (Weaver and Forcella 1986), its capacity to maintain/reproduce itself, to sustain itself, seems slight. That is, while seedlings appear in young stands (<300 years), they apparently do not survive for long. Both seedlings and small trees are essentially absent in older stands. The absence is likely due, in both cases, to subterranean competition with established trees. Fourth, while the stand may survive for >600 years, the sites must eventually be reset in two steps. Step 1 is by catastrophe, most likely fire—as indicated by plentiful charcoal in the soils of most or all WBP/whortleberry forests. Step 2 is by delivery of short-lived seeds from afar, i.e. they are not from 'in-house' seed banks. That delivery is surely from nearby productive stands by Clark's nutcracker.

ENVIRONMENT AND WBP

WBP often forms a timberline community. Its environment has been characterized with in-stand measurements in Montana (Weaver and Dale 1974) and across its north- south range (Weaver 1990a) to guide studies of its function, compared with vegetation zones above and below to explain its altitudinal distribution (Weaver 1977, 1979, 1980, 1994a), and compared with the environments of the five closely related stone pine species to demonstrate the similarity of the environmental requirements of all the stone pines (Weaver 1994a). Given easy access to this information via DOI and Scholar works references provided in the literature cited section, I focus instead on in situ community function, altitudinal distribution, and the within-stand competition leading to the low canopy cover normal to WBP stands. In comparing environment and exotic plant changes along the altitudinal gradient, I often substitute for WBP, qualities of a slightly lower altitude forest type, subalpine fir (*Abies lasiocarpa*). In the 'human influences' section, I consider whether the nearly identical climates of the five stone pines (Weaver 1994a) might allow the establishment of these species in the Rockies.

In Situ Function

WBP performs 'normally' in its normal habitat type (HT). We expect diameter growth of WBP to occur between May and September (bark slip), probably with a maximum (spring wood) in early summer when water, nutrients and temperatures are most favorable and tapering (summer wood), perhaps after cessation of branch growth, as water availability dwindles. Buds break in June, shoots grow in June-July and cones mature in September (Schmidt and Lotan 1980). I consider specific effects of environmental factors below, one at a time: temperature, water, nutrients, and competition.

WBP seedlings germinate and grow in the 10-40°C range, photosynthesis may be significant between 5-35°C (Jacobs and Weaver 1990) and these temperature ranges may suggest the physiological abilities of mature material. For example, photosynthesis probably occurs outside the diameter growth season, i.e. at least in April and in October whenever average maximum temperatures rise to 5-10°C and even intermittently during the winter. In general, temperatures measured in-stand correlate directly with temperatures experienced (e.g. air, soil and plant temperatures), thus temperature measurements correlate well with processes affected and so provide a solid basis for modeling.

In the northern Rocky Mountain summer, the subalpine fir/WBP HTs experience drought, at least in the 0-25 cm root zone. For instance, Weaver (1977) reported summer time water stress of >1 MPa in 1971 and >0.2 MPa in 1972. Drought was significantly greater in the Douglas fir zone below. In winter, soils under deep snow do not freeze. Thus, soil water is constantly replenished by snow melting from the snowpack above and strongly so as the pack begins to deteriorate in spring, perhaps as early as March (Weaver 1994a). Modelers should consider that in contrast to temperature data, in high mountain environments precipitation data are less dependable and may even be misleading. This is because much of apparently large water deposits (table 2) may be blown/sublimed off-site, intercepted and evaporated from trees or soils, run off, or percolate downward below the reach of roots—due to the low water holding capacity of the soils (table 2). Rare in-stand seasonal measures of soil water (e.g. Weaver 1977, Sirucek 1996) are needed, at least for confirmation of models.

Nutrient availability also varies with season (Weaver and Forcella 1979)—nutrients likely being released by decomposition from organic matter under the winter snowpack and being recaptured by spring-summer growth. Such nitrogen and phosphorus dynamics are demonstrated under experimental snowpacks in mountain meadows environmentally like those often adjacent to WBP stands (Weaver 1974, Weaver and Collins 1977, Yano et al. 2015).

If WBP were gardened, i.e. if competition were removed, it would likely be more successful in warmer and richer sites downslope. The escape of lodgepole pine in Argentina-Chile and Monterey pine (*Pinus radiata*) in Hawaii demonstrate such release. We discuss examples of such WBP release below.

Altitudinal Distribution

On an altitudinal transect, WBP is usually absent from dry grasslands of the plains (*Bouteloua gracilis*) and foothills (*Fesuca idahoensis*) and low forests (Douglas fir) above. It may appear in the subalpine fir zone, gains dominance upward toward timberline, and disappears in the alpine (Arno and Weaver 1990). The subalpine fir/alpine segment of this transect is beautifully demonstrated at the Big Sky Ski Resort near Bozeman, Montana (Weaver 1990b). Thus, WBP is a timberline species declining both upward and downward.

Temperature seems to have little direct effect on WBP's distribution, because average growing season temperature is near 10°C, which is within the seedling growth/photosynthesis range, minimum temperatures occur in the hardened period in winter, average July maxima are near seedling optima, and even absolute maximum temperatures allow photosynthesis and root growth (Jacobs and Weaver 1990, Weaver 1994a).

I suspect WBP is relatively drought intolerant, and so imagine that it may be inhibited upward by westerly winds in summer delivering dry air from lower altitudes or sand/ice scouring particles that reduce its tolerance by removing its needles' water-tight cuticles (Weaver 2001). In contrast to its exclusion from moist subalpine ridges by drought, at dry lower timberline (e.g. west slope of Montana's Crazy Mountains) one can see substantial WBP forest heavy with *Letharia vulpina* (fruticose lichen). I speculate that it occupies here a humid microenvironment formed by drainage of moist air from above and pooling behind glacial moraines. Counter-intuitively, its atmospheric demands seem to be bracketed by dry air in the moist climate above and moist air in the dry climate below.

Growing season length may also limit its distribution upward. The WBP community's productivity was related to the productivities of other communities on the Rocky Mountain altitudinal gradient by regressing their yields against various presumptive predictors. The best predictor by far (r^2 =0.85) was growing season length, defined as months with average air temperatures >0°C and no drought (Weaver 1994a). Thus, along the altitudinal gradient, production is low in high altitude sites (e.g. alpine and WBP) due to the short, low-temperature regulated season. Production is also low in grasslands due to the short, dry-soil regulated season, and greatest near the forest boundary where neither temperature nor water conditions are so extreme.

I attribute the diminishment of WBP downslope to competition with other conifers usually assumed to be favored by higher temperatures but which may be favored as much or more by soils that store/supply more water and nutrients (table 2). This hypothesis is supported by the fact that WBP can establish at these sites when its competitors are removed. That is, WBP appears on lower sites where fire or logging have removed competing conifers , especially so when understories established before the disturbance resist the success of small randomly distributed windblown conifer seed more than well provisioned WBP seeds planted in relatively plush open spots by Clark's nutcracker.

Still lower, grasslands and shrublands, perhaps especially fescue (*Festuca idahoensis*) grasslands, may be colonized by other conifers as well as WBP. Colonization may occur, especially if destructive fire is excluded or soil disturbance, e.g., rodent burrowing, cattle grazing, reduce competition. While small- seeded conifers may randomly broadcast their seeds more widely, the large well provisioned seeds of WBP work well with Clark's nutcracker in delivering them to secure sites and planting them at depth where their reserves support root penetration to relatively moist, rich, and less competitive soil layers. Despite the successful entry of such seedlings, they rarely find resources to become more than scraggly clutter in grazing lands.

Soils, Competition and Canopy Cover

As emphasized above, WBP forest structure is exceptionally open, orchardlike. The stand's openness must be due to competition among the trees for a resource other than light, which is surely not limiting in such open stands. This resource limitation cannot be absolute because basal area continues to increase through time but must, nevertheless, act constantly across years. Water supply is a candidate critical variable, being renewed each winter but being exhausted in late summer of each year. Nutrient supply (e.g., N, Ca, etc.; table 2) behaves similarly, being drawn down/exhausted by spring growth (twig extension and spring wood) and rejuvenated by decomposition under snow. The nutrient dynamic was examined across major HTs by Weaver and Forcella (1979).

Either or both limitations are related to soil properties. First, the low water holding capacity of the soil (table 2) provides little absolute buffer against summer drought—characteristic of the northern Rocky Mountains—and less buffering than in the downslope forests with greater canopy cover. Second, the large excess of precipitation over water holding capacity may lead to confusion regarding the position of WBP on the water gradient, as much of the precipitation goes unused. Third, the large excess of water results in leaching, which explains the relatively low concentrations of calcium, potassium, magnesium, and probably nitrogen in WBP soils, deficiencies that differentiate WBP from lower forests.

HUMAN IMPACTS

Pine nuts were a major foodstuff for Paleo-Indians and still are important to competing chipmunks, squirrels, Clark's nutcracker, and bears (*Ursus* spp.). Seed production values often are 20-250 m^{-2} or 2-25 gm^{-2} (Weaver and Forcella 1986). Production is variable and, though masting is significant, it is not regional. Masting favors highly mobile seed-consuming animals like nutcrackers over more sedentary mammals (Weaver and Forcella 1986). While nutcrackers harvest seeds from the treetops, squirrels often drop whole cones and cache them for winter use by themselves and bears, which again leads to less stand regeneration than dispersal by birds. The influence of animals on WBP establishment from seed was elegantly observed by Hutchins (1990) and tested experimentally (McCaughey and Weaver 1990a&b).

Other foodstuffs include whortleberries, which can be numerous, ranging 7-372 berries m^{-2}, the numbers not related to canopy cover (Weaver et al. 1990). Of the herbs present, glacier lily likely is the most nutritious while others (e.g. *Carex geyeri*) provide forage for cattle and wildlife (Weaver and Dale 1974, Forcella 1978, Forcella and Weaver 1979). Macro-fungal diversity may be similar to that in subalpine fir (~60 species and 0-163 kg ha-1) and Douglas fir forests (~61 species and 0- 216 kg ha^{-1}, Keck 2001).

Hiking trails in WBP forests cover a small area, damages mostly brittle shrubs and forbs, and introduce exotic species that remain at the trailside (Dale and Weaver 1974). Trampling damages or eliminates whortleberry, soils are strongly compacted, and both recover very slowly (Weaver and Dale 1978, Weaver et al. 1979). Trampling impacts at these sites increased from hikers to horses; motorcycles were most damaging of all going uphill and least damaging going downhill (Weaver and Dale 1978). The study sites from which these conclusions were drawn could be resampled, with little effort, for a unique 50-year reanalysis.

If burned, WBP forests are likely to return if seeds are locally available and nutcrackers deliver and plant them. The same might occur if sites were lightly logged. If logged, exotic species surely will increase, relatively scarce nutrient elements (table 2) will be exported, and compaction will occur. Nutrients are most concentrated in needles, twigs and bark. Thus, the amount of exported nutrients declines from harvest of whole trees, through boles, to least with peeled boles, as is done in some Italian forests (Weaver and Forcella 1977).

Invasion of exotic species in WBP and fifteen other environmental zones was compared by recording their presence in four disturbance zones of each (table 3, Weaver et al. 1990, 1995, 2001) In the WBP zone, 14 species with measurable cover (table 3, shaded) were found on constantly disturbed road-shoulders, six were found on road-cuts (once disturbed and undergoing primary succession), four were found on cleared right-of-way, where they competed with residual understory vegetation, i.e. secondary succession, and none appeared in undisturbed forest, where invaders compete with both over and understory vegetation of the 'near climax' vegetation originally present. Similar trends occurred where the responding variable was total exotic species present (15, 12, 12, 0 species) or exotic species with >30% constancy (13, 11, 6, 0 species, table 3). The principal exotic species were domestic grasses (*Bromus inermis*, *Poa pratensis*, and *Phleum pratense*), dandelion (*Taraxacum officinale*), and legumes (*Trifolium hybridum*, *T. repens*, *Melilotus officinale*, and *Medicago lupulina*).

With respect to exotic species, the generality and application of the WBP conclusions can be compared with observations in four other vegetation zones (table 3). The total number of exotic species increases from alpine (6) and WBP (15) to forests and grasslands below (18-21). We suggest that the colder environments above are probably too rigorous for exotics normally arriving from warmer sites or that WBP sites may be too remote for propagule delivery. In all zones, the number of exotic species declines with decreased disturbance and increased native competition, although the decline weakens downslope. Note that data from cleared rights-of-way are lacking in alpine and grassland sites in table 3 because no trees were

Table 3. Exotic species presence/numbers by habitat type (HT) and disturbance regime.

		All exotic species[e]				Number of exotic species[c] Constancy >30%[e]				Species with measurable cover[f]			
HT[a]	reps	Sh[b]	Cut	Clear	UnD	Sh[b]	Cut	Clear	UnD	Sh[b]	Cut	Clear	UnD
Alpine	11	6	4		4	2	1		0	1	0		0
WBP	10	15	12	12	0	13	11	6	0	14	6	4	0
Psme	10	20	17	12	9	11	9	5	5	10	7	4	3
Agsp	8	18	18		15	9	11		8	8	4		4
Bogr	7	21	21		19	12	12		9	7	5		1

[a]Environmental zones (HTS) are Alpine, forest (WBP/*Abies lasiocarpa* and Psme/*Pseudotsuga menziesii*) grassland (moist, *Agropyron spicatum* Agsp and dry, *Bouteloua gracilis* Bogr)

[b]Disturbance regimes are road shoulder (Sh), road cut (Cut, 1°), cleared (Clear, 2°) and undisturbed (UnD)

[c]Exotic species number is reported with three measures.

[d]number of exotic species even if only in one replication

[e]number of exotic species in >30% of the stands

[f]number of exotic species with measurable cover

cleared from these zones. These observations can be applied to broader two-dimensional landscapes of forest and rangeland. If the establishment of exotic species is proportional to the amount of competition/disturbance, then establishment of exotic species on disturbed sites near and far from roads should be equal, i.e. exotic species establishment in logged areas (far distance) might be similar to that of cleared rights-of-way (near distance). This assumption may over-estimate the invasion of remote sites because delivery of propagules is less there than along trails or roads.

Introduction of the white pine blister rust from Europe has been, by far, man's most damaging act for WBP. I suggested in 1998 that WBP's extreme susceptibility might be greatly reduced by introducing genes from closely related and highly resistant European stone pines. Their resistance may well be transferred by simple hybridization, i.e. collect European pollen (*Pinus cembra* or *P. sibirica*) and transfer it to mature WBP in wild or arboretum settings. Success is probable because closely related pines usually hybridize easily. Thus, I expect the F1 to be polygenically resistant and, as the F1s backcross in the wild, some F2 will be even more resistant (naturally selected) and some less resistant (naturally eliminated). Potential difficulties include: 1) Species may be genetically incompatible despite the common tendency of closely related pines to hybridize, 2) Despite similar latitudes, phenological (day-length) differences may inhibit natural crossing and backcrossing. If so, the first crosses must be artificial but the failure will fade as backcrossing dilutes any Eurasian daylength control, 3) F1 and F2 progeny might not tolerate the WBP environment. This is unlikely since half their genes are native and half come from trees drawn from similar environments (Weaver 1994b), 4) Genes introduced with the resistant European material explode to create a weedy invader. This also is unlikely since the F1 are immediately half WBP and in each succeeding generation, the European genes will become further diluted. The probability of concentration of European genes is diminishingly small, first, because pollen from surrounding native WBP is much more available than pollen from the F1 and F2 hybrid trees and, second, selection for European genes will be low because European phenotypes are doubtless less well adapted to Rocky Mountain than Swiss environments, 5) Immediate introduction might have allowed natural crossing and free spread of the resistance genes in the wild (as suggested by parenthetic 'in the wild' above). While cautious controlled breeding, might be favored by some, this approach would be more expensive in dollars and time. Currently, lab/garden crossing may be necessary anyway, because natural populations of WBP are already so sparse.

ACKNOWLEGEMENTS

I review, integrate, and extend understanding of WBP, its environment and human impacts in the forests hopefully to good advantage. My work depends on the direct contributions of those listed in the introduction, many associates, and partial funding by USFS, NSF, NPS. I also thank my reviewers, F Forcella, L Rew, and J Habeck.

LITERATURE CITED

Arno, S, and T Weaver. 1990. Whitebark pine community types and their patterns on the landscape. In: W Schmidt and K McDonald, eds. Symposium on whitebark pine ecosystems: ecology and management of a high mountain resource. USDA Forest Service General Technical Report INT-270:97-105. Ogden UT. doi.org/10.2737/INT-GTR-270.

Dale, D, and T Weaver. 1974. Trampling effects on vegetation of the trail corridors of northern Rocky Mountain forests. Journal of Applied Ecology 11:767-772. doi.org/10.2307/2402226. https://scholarworks.montana.edu/xmlui/handle/1/16434.

Eversman, S, C Johnson, and D Gustafson. 1990. Vertical distribution of epiphytic lichens on three tree species in Yellowstone National Park. In: W Schmidt and K McDonald, eds. Symposium on whitebark pine ecosystems: ecology and management of a high mountain resource. USDA Forest Service General Technical Report INT-270:367-368. Ogden UT. doi.org/10.2737/INT-GTR-270.

Forcella, F. 1978. Flora and chorology of the *Pinus albicaulis-Vaccinium scoparium* association. Madroño 25:139-150.

Forcella, F, and T Weaver. 1977. Biomass and productivity of the subalpine *Pinus albicaulis-Vaccinium scoparium* association in Montana, USA. Vegetation 35:95-105.

Forcella, F, and T Weaver. 1979. Food production in the *Pinus albicaulis/ Vaccinium scoparium* association. Montana Academy of Sciences Proceedings 39:73-80.

Hutchins, H. 1990. Whitebark pine seed dispersal and establishment: who's responsible? In: W Schmidt and K McDonald, eds. Symposium on whitebark pine ecosystems: ecology and management of a high mountain resource.

USDA Forest Service General Technical Report INT-270:245-255. Ogden UT. doi.org/10.2737/INT-GTR-270.

Jacobs, J, and T Weaver. 1990. Effects of temperature and temperature preconditioning on seedling performance of whitebark pine. In: W Schmidt and K McDonald, eds. Symposium on whitebark pine ecosystems: ecology and management of a high mountain resource. USDA Forest Service General Technical Report INT-270:134-139. Ogden UT. doi.org/10.2737/INT-GTR-270.

Keane, R, S Arno, J Brown, and D Tombach. 1990. Simulating disturbances and conifer succession in whitebark pine forests. In: W Schmidt and K McDonald, eds. Symposium on whitebark pine ecosystems: ecology and management of a high mountain resource. USDA Forest Service General Technical Report INT-270:274-298. Ogden UT. doi.org/10.2737/INT-GTR-270.

Keck, J. 2001. Macrofungi of the altitudinal gradient, Northern Rocky Mountains. MS thesis, Montana State University, Bozeman MT. 73 pp. doi.org/10.15788/8101.

McCaughey, W, and T Weaver. 1990a. Biotic and microsite factors affecting whitebark pine establishment. In: W Schmidt and K McDonald, eds. Symposium on whitebark pine ecosystems: ecology and management of a high mountain resource. USDA Forest Service General Technical Report INT-270:140-150. Ogden UT. doi.org/10.2737/INT-GTR-270.

McCaughey, W, and T Weaver. 1990b. Reference guide to whitebark pine, *Pinus albicaulis* Engelmann. In: W Schmidt and K McDonald, eds. Symposium on whitebark pine ecosystems: ecology and management of a high mountain resource. USDA Forest Service General Technical Report INT-270:376-386. Ogden UT. doi.org/10.2737/INT-GTR-270.

Pfister, R, B Kovalchik, S Arno, and R Presby. 1977. Forest habitat types of Montana. USDA Forest Service General Technical Report INT-34:174 pp. Ogden UT.

Schmidt, WC, and J Lotan. 1980. Phenology of common forest flora of the northern Rockies: 1928-1937. USDA Forest Service Research Paper INT-259. Ogden UT.

Sirucek, D. 1996. Soil temperature and soil moisture characteristics for several habitat types of Montana and Idaho. MS thesis, Montana State University, Bozeman MT. doi.org/10.15788/7452-1996.

Weaver, T. 1974. Ecological effects of weather modification: effect of late snow melt on *Festuca idahoensis* meadows. American Midland Naturalist 92:346-356. doi.org/10.2307/2424300. https://scholarworks.montana.edu/xmlui/handle/1/16422.

Weaver, T. 1977. Root distribution and soil water regimes in nine habitat types of the northern Rocky Mountains. In: J Marshall. The belowground ecosystem. Range Science Series 26:239-244. Colorado State University, Fort Collins CO. https://scholarworks.montana.edu/xmlui/handle/1/16566.

Weaver, T. 1979. Changes in soils along a vegetational (altitudinal) gradient of the northern Rocky Mountains. In: C Youngberg, ed. Proceedings of the Fifth North American Forest Soils Conference, pp 14-29. Soil Science Society of America, Madison WI. https://scholarworks.montana.edu/xmlui/handle/1/16567.

Weaver, T. 1980. Climates of vegetation types of the northern Rocky Mountains and adjacent plains. American Midland Naturalist 103:392-398.

Weaver, T. 1990a. Climates of subalpine pine woodlands. In: W Schmidt and K McDonald, eds. Symposium on whitebark pine ecosystems: ecology and management of a high mountain resource. USDA Forest Service General Technical Report INT-270:72-79. Ogden UT. doi.org/10.2737/INT-GTR-270.

Weaver, T. 1990b. Seeing whitebark pine in a northern Rocky Mountain (USA) landscape: notes for a field trip. In: W Schmidt and K McDonald, eds. Symposium on whitebark pine ecosystems: ecology and management of a high mountain resource. USDA Forest Service General Technical Report INT-270:355-358. Ogden UT. doi.org/10.2737/INT-GTR-270.

Weaver, T. 1994a. Climates where stone pines grow, a comparison. In: W Schmidt and K Holtmeier, eds. Stone pines and their environments: the status of our knowledge. USDA Forest Service General Technical Report INT-GTR-309:85-89.

Weaver, T. 1994b. Vegetation distribution and production in Rocky Mountain climates — with emphasis on whitebark pine. In: W Schmidt and K Holtmeier, eds. Stone pines and their environments: the status of our knowledge.

USDA Forest Service General Technical Report INT-GTR-309:142-152.

Weaver, T. 2001. Whitebark pine in its environment. In: D Tomback, S Arno, and R Keane, eds. Whitebark pine communities: ecology and restoration. Island Press, Washington DC. 41-73.

Weaver, T. 2022. WBP community processes, environment and human influences: revisiting MSU work of 1971-2000, extended version. Unpublished paper. Ecology Department, Montana State University, Bozeman MT.

Weaver, T, and D Collins. 1977. Possible effects of weather modification (increased snow pack) on *Festuca idahoensis* meadows. Journal of Range Management 30:451-456.

Weaver, T, and D Dale. 1974. *Pinus albicaulis* in central Montana: environment, vegetation and production. American Midland Naturalist 92: 222-230. doi.org/10.15788/16400-1974.

Weaver, T, and D Dale. 1978. Trampling effects of hikers, motorcycles and horses in meadows and forests. Journal of Applied Ecology 15:451-457. doi.org/10.2307/2402226.

Weaver, T, D Dale, and E Hartley. 1979. The relationship of trail condition to use, vegetation, user, slope, season and time. In: R Ittner, D Potter, J Agee and S Anscheil, eds. Recreational impact on wildlands. USDA Forest Service Pacific Northwest Region #R-6-001-1979:94-100.

Weaver, T, and F Forcella. 1977. Biomass of fifty conifer forests and nutrient exports associated with their harvest. Great Basin Naturalist 37:395-401.

Weaver, T, and F Forcella. 1979. Seasonal variation in soil nutrients under six Rocky Mountain vegetation types. Soil Science Society of America 43:589-593. https://digitalcommons.usu.edu/aspen_bib/4588/.

Weaver, T, and F Forcella. 1986. Cone production in *Pinus albicaulis* forests. In: R Shearer, ed. Proceedings - Conifer tree seed in the inland mountain west symposium. USDA General Technical Report IOONT-203:68-76. Ogden UT. https://scholarworks.montana.edu/xmlui/handle/1/16565.

Weaver, T, F Forcella, and D Dale. 1990. Stand development in whitebark pine woodlands. In: W Schmidt and K McDonald, eds. Symposium on whitebark pine ecosystems: ecology and management of a high mountain resource. USDA Forest Service General Technical Report INT-270:151-155. Ogden UT. doi.org/10.2737/INT-GTR-270.

Weaver, T, D Gustafson, and J Lichthardt. 1995. Plants establishing in Rocky Mountain environments — a manual for choosing native species for revegetation. MSU Biology Report # 42:71 pp. Montana State University, Bozeman MT. https://scholarworks.montana.edu/xmlui/handle/1/16568.

Weaver, T, D Gustafson, and J Lichthardt. 2001. Exotic plants in early and late seral vegetation of fifteen northern Rocky Mountain environments (HTs). Western North America Naturalist 61:417-427.

Weaver, T, and J Jacobs. 1990. Occurrence of multiple stems in whitebark pine. In: W Schmidt and K McDonald, eds. Symposium on whitebark pine ecosystems: ecology and management of a high mountain resource. USDA Forest Service General Technical Report INT-270:156-159. Ogden UT. doi.org/10.2737/INT-GTR-270.

Weaver, T, K Kendall, and F Forcella. 1990. Berry production in three whitebark pine forest types. In: W Schmidt and K McDonald, eds. Symposium on whitebark pine ecosystems: ecology and management of a high mountain resource. USDA Forest Service General Technical Report INT-270:198-200. Ogden UT. doi.org/10.2737/INT-GTR-270.

Weaver, T, J Lichthardt, and D Gustafson. 1990. Exotic invasion of timberline vegetation, northern Rocky Mountains USA. In: W Schmidt and K McDonald, eds. Symposium on whitebark pine ecosystems: ecology and management of a high mountain resource. USDA Forest Service General Technical Report INT-270:208-213. Ogden UT. doi.org/10.2737/INT-GTR-270.

Yano, Y, ENJ Brookshire, JH Holsinger, and T Weaver. 2015. Long-term snowpack manipulation promotes large loss of bioavailable nitrogen and phosphorus in a subalpine grassland. Biogeochemistry 124:319-333. doi:10.1007/s10533-015-0100-9. http://link.springer.com/article/10.1007/s10533-015-0100-9.

Maintaining *Pinus Strobiformis*, a Tree Species Threatened by Climate Change and White Pine Blister Rust

Nicholas Wilhelmi, <nicholas.wilhelmi@usda.gov>, USDA Forest Service, Forest Health Protection, Flagstaff, Arizona USA; **Kristen M. Waring**, School of Forestry, Northern Arizona University, Flagstaff, Arizona USA; **Richard A. Sniezko**, USDA Forest Service, Dorena Genetic Resource Center, Cottage Grove, Oregon USA; **Gregory J. Reynolds**, USDA Forest Service, Forest Health Protection, Albuquerque, New Mexico USA; and **Jeremy S. Johnson**, School of Forestry, Northern Arizona University, Flagstaff, Arizona USA; USDA Forest Service, Dorena Genetic Resource Center, Cottage Grove, Oregon USA and Department of Environmental Studies, Prescott College, Prescott, Arizona USA.

ABSTRACT

Climate change and the non-native disease white pine blister rust (WPBR) pose significant threats to the health and resilience of southwestern white pine (SWWP; *Pinus strobiformis*), native to the southwestern US and Mexico. Seed sources with adaptive traits needed to survive warmer, drier conditions and durable genetic resistance to WPBR are critical to maintaining this important species on the landscape. To facilitate this need, seedlings from across the range of SWWP were tested for resistance to WPBR at Dorena Genetic Resource Center in Oregon. Major gene and quantitative resistance were documented. Data were also used to estimate resistance levels and frequency across the range of SWWP. Scion from parent trees identified as resistant was collected and grafted into a clone bank orchard at John T. Harrington Forestry Research Center in Mora, NM. In addition, progeny identified as resistant will be grafted into a clone bank orchard at Tyrell Seed Orchard in Oregon to maintain these genetics. Seed will be collected from these orchards and resistant parent trees will be planted on the landscape. We have also established two common-garden field trials in the Southwest to validate resistance results, monitor the long-term durability of resistance, and assess adaptive traits. This work provides information critical to identifying seed sources for future planting. Current challenges include funding ongoing activities (testing, outplanting and monitoring), variable and unpredictably dry planting conditions, and mortality of original parent trees. This interdisciplinary, collaborative project includes international, academic, federal, and tribal partners.

INTRODUCTION

Climate change and invasive pathogens pose significant threats to forest health and resilience. Southwestern white pine (SWWP, *Pinus strobiformis*), native to the southwestern USA and Mexico, faces increasing pressure from hotter, drier conditions and white pine blister rust (WPBR), a disease caused by the non-native fungal pathogen *Cronartium ribicola*. Seed sources with durable genetic resistance to WPBR and the adaptive traits needed to survive warmer, drier conditions in the future are essential to sustain this species on the landscape. Inoculation trials in collaboration with Dorena Genetic Resource Center (DGRC) have shown SWWP to be highly susceptible to WPBR. However, moderately high resistance has been identified in a low percentage of families (Johnson and Sniezko 2021), with much more extensive

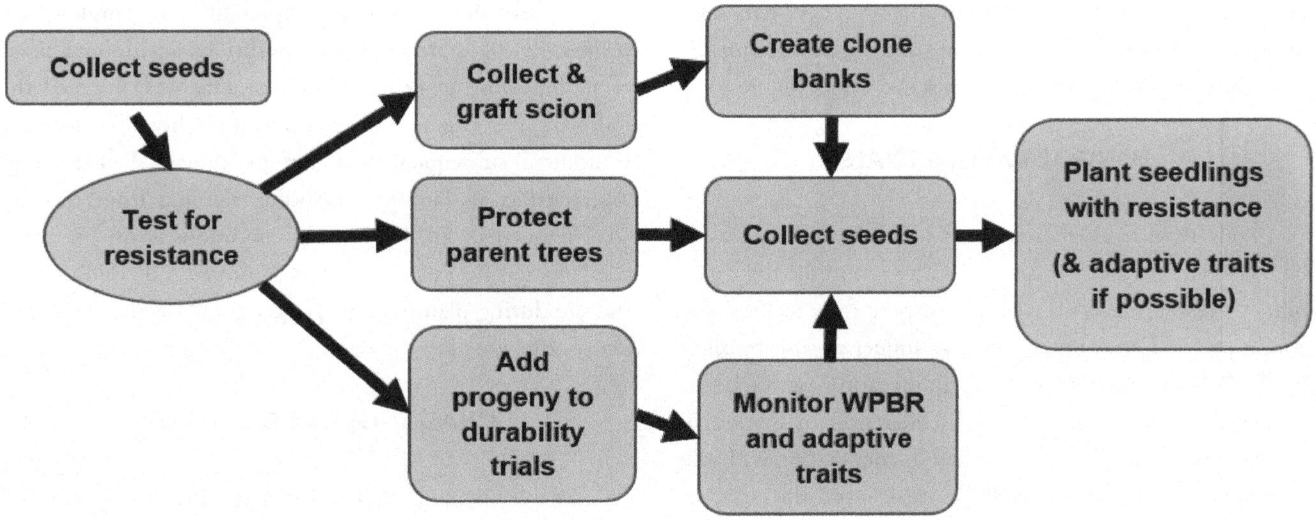

Figure 1. Flow chart demonstrating our multifaceted approach to identify and conserve seed sources with desirable adaptive traits such as drought tolerance and resistance to white pine blister rust.

testing currently in progress. We are using a multi-faceted approach to develop a collection of seed sources containing genetic resistance to WPBR and adaptive traits, such as drought tolerance, that can be used to enhance operational plantings of SWWP and increase long-term resilience in Southwest US populations (figure 1). Collaboration between Northern Arizona University (NAU), Forest Health Protection (FHP), and DRGC has produced valuable range-wide information on resistance and adaptive traits in a relatively short time span compared to other species.

SCION COLLECTIONS AND CLONE BANK

Seedlings from across the range of SWWP are currently in testing for resistance to WPBR at DGRC in Oregon. In the first trials, both major gene resistance (MGR) and quantitative resistance (QR) were documented (Johnson and Sniezko 2021; Sniezko et al. 2008; Waring et al. 2019). Scion material was collected from parent trees whose progeny were identified at DGRC as having MGR or QR to WPBR (figure 2A). Scion collections have thus far occurred in New Mexico on the Sacramento Mountains, Lincoln National Forest (2015, 2016, 2018) and Mescalero Apache Tribal Lands (2020); Zuni Mountains, Cibola National Forest (2018, 2019); and in Arizona on the Apache-Sitgreaves National Forests (2019, 2020). Tree climbers collected scion from branches in the upper third of resistant tree crowns to maximize future seed production. Following collection, which takes place in the winter, scion material was packed in moistened sphagnum moss and stored in a walk-in cooler until spring. It was then grafted onto rootstock that originated from a bulked SWWP seedlot (figure 2B) and eventually outplanted in a clone bank at the John T. Harrington Forestry Research Center in Mora, NM (a New Mexico State University facility) (figure 2C). This clone bank will preserve parental genotypes for WPBR resistance in SWWP, even if parent trees are lost to the increasingly high risk of wildfire. Collections will continue as data are analyzed and new resistant parent trees are identified. Resistant seed collected from parent trees in the field or the clone bank will eventually be provided to various partners.

PROGENY CLONE BANK AT TYRELL SEED ORCHARD

Progeny, including those which have shown resistance, are generally removed to make room for active trials following inoculation and resistance assessments at DGRC. Region 3 Forest Health Protection and NAU are collaborating with DGRC and the Bureau of Land Management (BLM) to establish a clone bank/seed orchard at BLM's Tyrell Seed Orchard to preserve progeny that have demonstrated resistance. Establishing this clone bank will enable conservation of these valuable genotypes without transporting material inoculated in Oregon to the Southwest and risking the introduction of new races of the WPBR pathogen. To date, 102 grafts representing 38 families from sow years 2002 and 2009 have been grafted and are scheduled to be planted in early 2022. Grafting will continue as trials are completed for sow years 2014,

2016, 2017, and 2020. All individuals grafted remained healthy throughout the inoculation trials, demonstrating either complete (MGR) or partial resistance (QR) (figure 3).

LONG-TERM FIELD TRIALS

Long-term field trials are used to further validate seedling assessment results from DRGC and monitor the durability of resistance. Field trials also expose trees to local site conditions and stressors which may influence susceptibility to WPBR and provide valuable information on variation in adaptive traits, such as drought tolerance. In Region 3, two field trials have been established in cooperation with the Mescalero Apache (2017) in New Mexico and the Apache-Sitgreaves National Forests (2018) in Arizona (figure 4). Both sites were established in areas currently infected by WPBR. However, no WPBR infection has been observed on either site thus far.

In Arizona, 1,092 seedlings originating from 21 open-pollinated half-sib families were planted in 2019. Due largely to hot and dry planting conditions, 40% mortality was observed the year following planting (figure 5). A second round of planting, consisting of 1,200 trees representing 57 families, including families from Mexico, was completed amid much more favorable conditions in 2021.

In New Mexico, 231 trees representing 61 families were planted in 2017. Due to harsh conditions, including a flood event following planting, 37% mortality was observed the following year (figure 6). Significant mortality has continued to occur in subsequent years. An additional 1,160 seedlings representing 56 families, including families from Mexico, were planted in 2021. Planting conditions again presented challenges as heavy rains created saturated soil conditions at the site during planting and pocket gopher activity was evident across much of the site.

CHALLENGES AND REWARDS

Establishing and maintaining a program to identify and preserve well adapted genotypes that are resistant to WPBR presents many challenges and rewards. Mortality of known resistant parent trees due to fire, insects/disease, and other abiotic stressors is a constant challenge and demonstrates the urgency and importance of scion and seed collection from these trees. Grafting has presented a challenge as many grafts are unsuccessful for various reasons, including difficulty locating scion of suitable size (around 10 mm minimum) on some parent trees. The long-term durability of field site plantings has presented many challenges, mainly related to weather conditions, which are increasingly unpredictable. In

Figure 2. A tree climber collecting scion material (A), grafted scion in the greenhouse (B), and an out-planted graft (C).

Figure 3. Surviving southwestern white pine seedlings in 2021 that had been inoculated with *Cronartium ribicola* in 2010 from the SY2009 trial at Dorena Genetic Resource Center (DGRC) (A) and grafts from the SY2009 trial in greenhouses (B) at DGRC; note the lack of disease symptoms in these resistant trees. Photos: Richard Sniezko (A), Emily Boes (B).

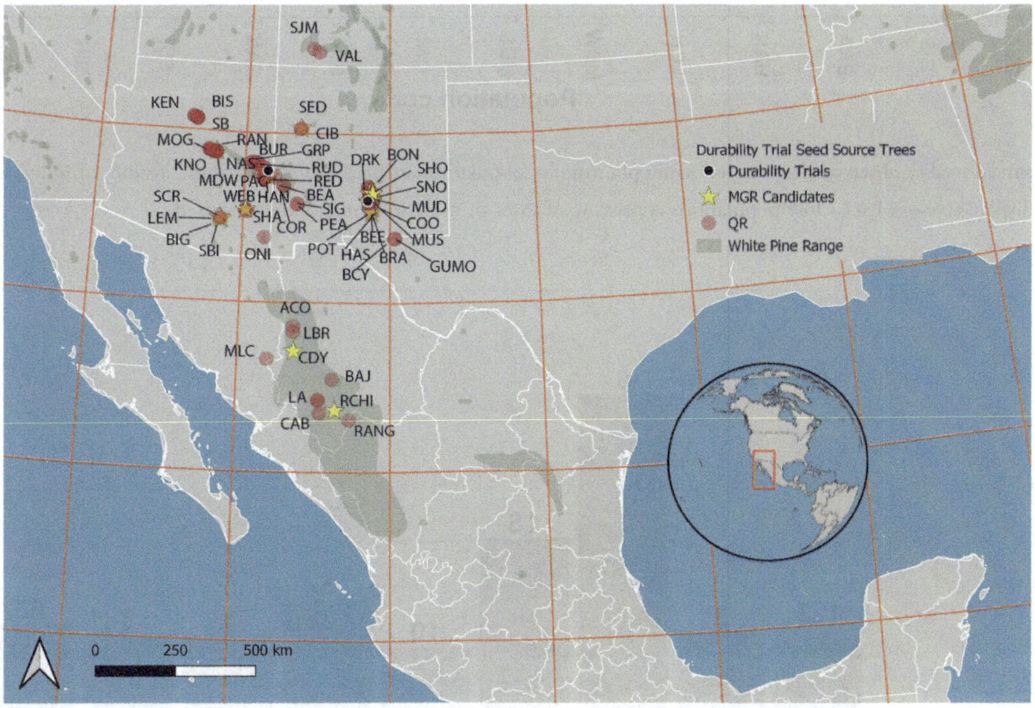

Figure 4. Locations of field sites and seed sources planted within the field sites.

the plantings thus far, we have experienced flooding, saturated soils, animal activity, and hot dry conditions, all of which have influenced survival. With challenges come great rewards, particularly the identification and conservation of the first documented WPBR-resistant genotypes of this species. This work is a collaboration between FHP, NAU, several national forests, and the Mescalero Apache Tribe (figure 7). This collaboration has provided great opportunities to work with, educate, and learn from all those involved in the project.

ACKNOWLEDGMENTS

We thank Angelia Kegley and the many technicians at DGRC who have helped with the assessments, especially Evan Heck, Brianna McTeague, Bob Danchok, Emily Boes, and Megan Lewien. Thanks to Al Hendricks for all the hard work building and planting the site and to the many other technicians from the Waring Lab at NAU. We thank Dave Conklin, former FHP pathologist, for initial seed collections and his work with WPBR in Region 3. Thank you to

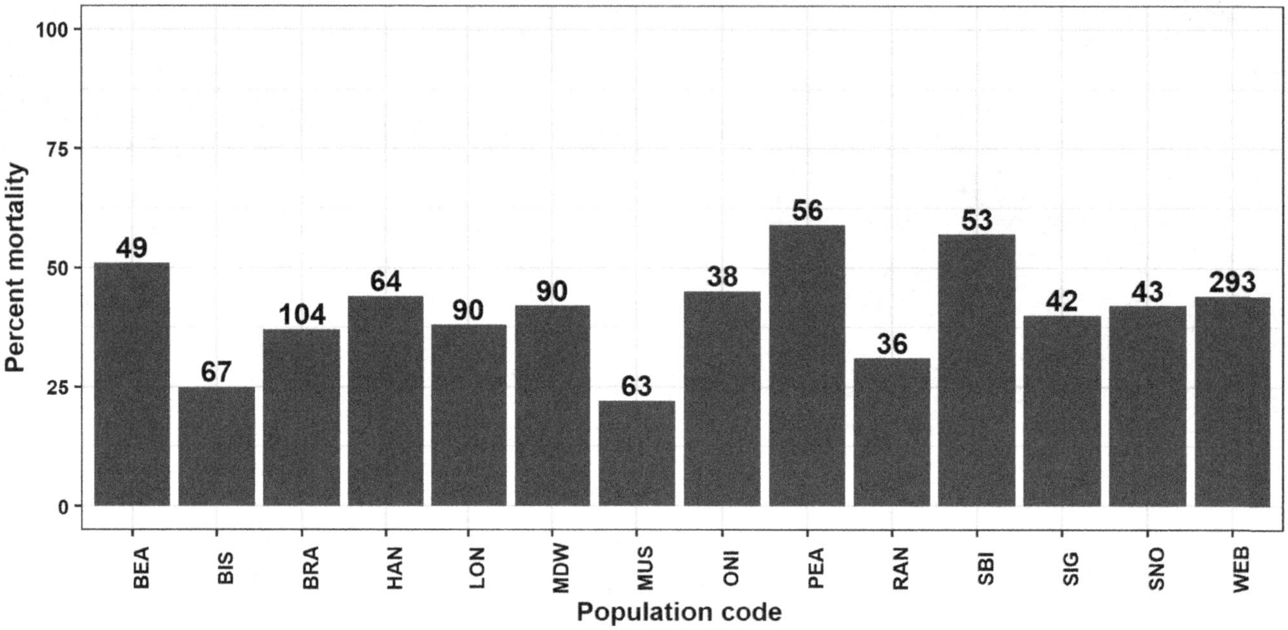

Figure 5. Percent mortality observed the year following planting by population at the Arizona field site. Number of individuals planted per population is displayed above bars. Only populations consisting of three or more individuals are presented.

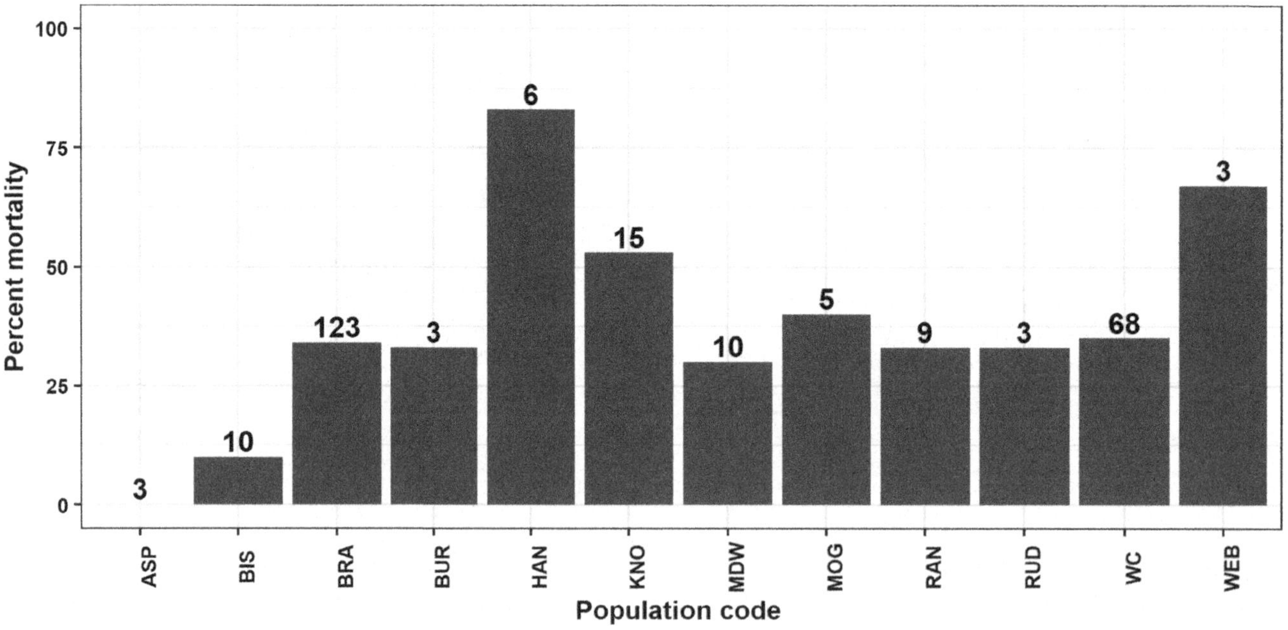

Figure 6. Percent mortality observed the year following planting by population at the New Mexico field site. Number of individuals planted per population is displayed above bars. Only populations consisting of three or more individuals are presented.

Figure 7. Planting of the field site located on Mescalero Apache Tribal Lands in 2021.

Christian Wehenkel of Universidad Juárez del Estado de Durango for the Mexican seed collections. Many thanks to Phil Smith, Thora Padilla, and the many technicians of the Mescalero Apache; Bill Hornsby of the Bureau of Indian Affairs; and Matt Harrison, Jacob Meulebroeck, and other Apache-Sitgreaves National Forests personnel. Without the cooperation and support of the Mescalero Apache, Bureau of Indian Affairs, and the Apache-Sitgreaves National Forests this work would not be possible. This material is based in part upon work supported by the National Science Foundation under Grant No. EF-1442597, the USDA Forest Service Region 6 Genetic Resource Program, Forest Health Protection, and the Gene Conservation Program.

LITERATURE CITED

Johnson, JS, and RA Sniezko. 2021. Quantitative disease resistance to white pine blister rust at southwestern white pine's (*Pinus strobiformis*) northern range. Frontiers in Forests and Global Change 4:765871. doi: 10.3389/ffgc.2021.765871.

Sniezko, RA, Kegley, AJ, and R Danchok. 2008. White pine blister rust resistance in North American, Asian, and European species—results from artificial inoculation trials in Oregon. Annals of Forest Research 51:53–66.

Waring, KM, S Cushman, A Eckert, L Flores-Renteria, R Sniezko, C Still, C Wehenkel, A Whipple, M Wing, J Bagley, E Bucholz, S Gonzalez-Elizondo, J Hartsell, C Garms, E Heck, J Johnson, E Landguth, A Leal-Saenz, M Menon, E Moler, G Page, J ShawJ, A Shirk, and J Swenson. 2019. Ecology and evolution of a tree species challenged by dual threats. XXV IUFRO World Congress. Sep 29-Oct 5 2019, Curitiba, Brazil.

Forest Structure Twenty Years After the First Whitebark Pine Prescribed Burn in Banff National Park

Brendan Wilson, <bwilson@selkirk.ca>, School of Environment & Geomatics, Selkirk College, BC, Canada; **Stephanie Jouvet**, School of Environment & Geomatics, Selkirk College, BC, Canada; **Jon Stuart-Smith**, Parks Canada Agency, Banff National Park, AB, Canada; **Rob Walker**, Retired, Parks Canada Agency, Kootenay National Park, BC, Canada.

ABSTRACT

Following a reconnaissance level survey of blister rust infection in the Canadian Rockies, Parks Canada started a conservation program for whitebark pine in 1998. With the recognition that fire suppression had reduced the amount of available early seral subalpine forest in the Parks system, the initial action was to assess the use of prescribed fire in creating suitable recruitment habitat for this species. Permanent monitoring plots were established in control and burn treatment units on the southern flank of Observation Peak in Banff National Park. After the initial pre-burn measurement of the treatment units, a high intensity fire was applied to the burn treatment in late August of 1998. The treatments were remeasured the following summer. Twenty years later, we found substantial recruitment of whitebark pine, compared to the spruce and fir species that previously dominated the burn treatment area. No blister rust was evident on whitebark pine in either treatment. Understory shrub layers were dominated by huckleberry and grouseberry in the burn treatment, whereas, the control was dominated by a more diverse mixture subalpine fir, common juniper, Engelmann spruce, and grouseberry.

Keywords: BACI design; prescribed fire; restoration; tree regeneration; understory response; permanent plots

INTRODUCTION

Whitebark pine (*Pinus albicaulis*) is an essential part of subalpine ecosystems in the Canadian Mountain National Parks. This high-elevation keystone species' seeds provide an important food source for a number of animals including squirrels, bears, and in particular, the bird species Clark's nutcracker (*Nucifraga columbiana*) (Smith et al. 2008). Following a reconnaissance level survey of whitebark pine blister rust infection in the Canadian Rockies (Stuart-Smith 1998), Parks Canada started a conservation program for whitebark pine (Wilson and Stuart-Smith 2002). Recognizing that fire suppression had reduced the amount of available early seral subalpine forest in the ecosystem, the initial action was to assess the use of prescribed fire in creating suitable recruitment habitat for this species, and quantitatively document the successional patterns over time.

METHODS

Forty100 m^2 permanent monitoring plots were established in control and burn treatment units on the southern flank of Observation Peak (Helen Ridge) in Banff National Park to create a Before and After Controlled Impact study design (figure 1). Initial pre-burn measurements of the treatment units recorded foliage cover, height, and vascular plant species composition of four forest structure layers (table 1). Standard tree species demographic records were also recorded including height, diameter at breast height, vigour, and disease status (B.C. Ministry of Forests and Range and BC Ministry of Environment 2010).

Table 1. Forest stratification layers and their respective height ranges.

Forest Layer	Height Range (cm)
Canopy (A)	>=10 m
Tall Shrub (B1)	>=2 m and <10 m
Low Shrub (B2)	>=15 cm and <2 m
Herb Layer (C)	Forbs, grasses, dwarf woody <15 cm

After the initial pre-burn measurement of the treatment units, a high intensity fire was applied to the burn treatment in late August of 1998. The treatment units were remeasured the following summer and the results confirmed that the fire had achieved almost 100% mortality of all size classes of the tree and shrub species present (Wilson et al. 2002).

RESULTS AND DISCUSSION

Twenty years later, we remeasured both the treatment and control plots and found substantial recruitment of whitebark pine, compared to the spruce and fir species that previously dominated the burn treatment plots (figure 2-3). The density of all tree species combined was 2200 stems/hectare in the burn treatment, with whitebark accounting for just under half of this value. Greater that 99% of all tree species were seedlings or saplings less than 1.5m tall. No blister rust was evident on whitebark pine in either treatment in the past or at this latest measurement time.

The dominant vegetation in the canopy layer in the unburned control unit remained overwhelmingly Englemann spruce (*Picea englemannii*) (~0.75) with subalpine fir (*Abies lasiocarpa*) making up the rest of the measured cover (figure 4). The remaining standing dead trees in the burned plots provided approximately 5% canopy cover in that treatment. Subalpine fir was the dominant species in the unburned control tall shrub layer (2 to 10 m), with some spruce and whitebark pine.

Understory shrub and herb layers were dominated by huckleberry (*Vaccinium membranaceum*) and grouseberry (*V. scoparium*) in the burn plots, whereas, the control plots were dominated by a more diverse mixture of subalpine fir, common juniper (*Juniperus communis*), Engelmann spruce, and grouse-

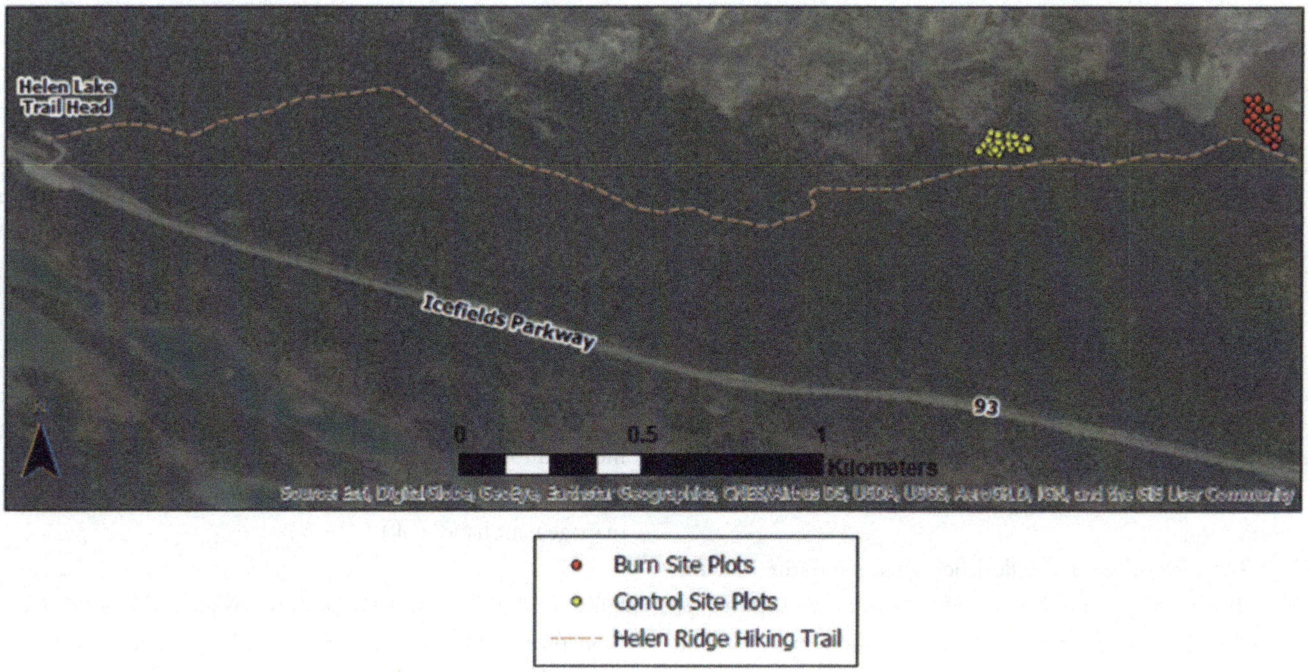

Figure 1. Study site location along the Helen Lake Hiking Trail, Banff National Park, AB, Canada.

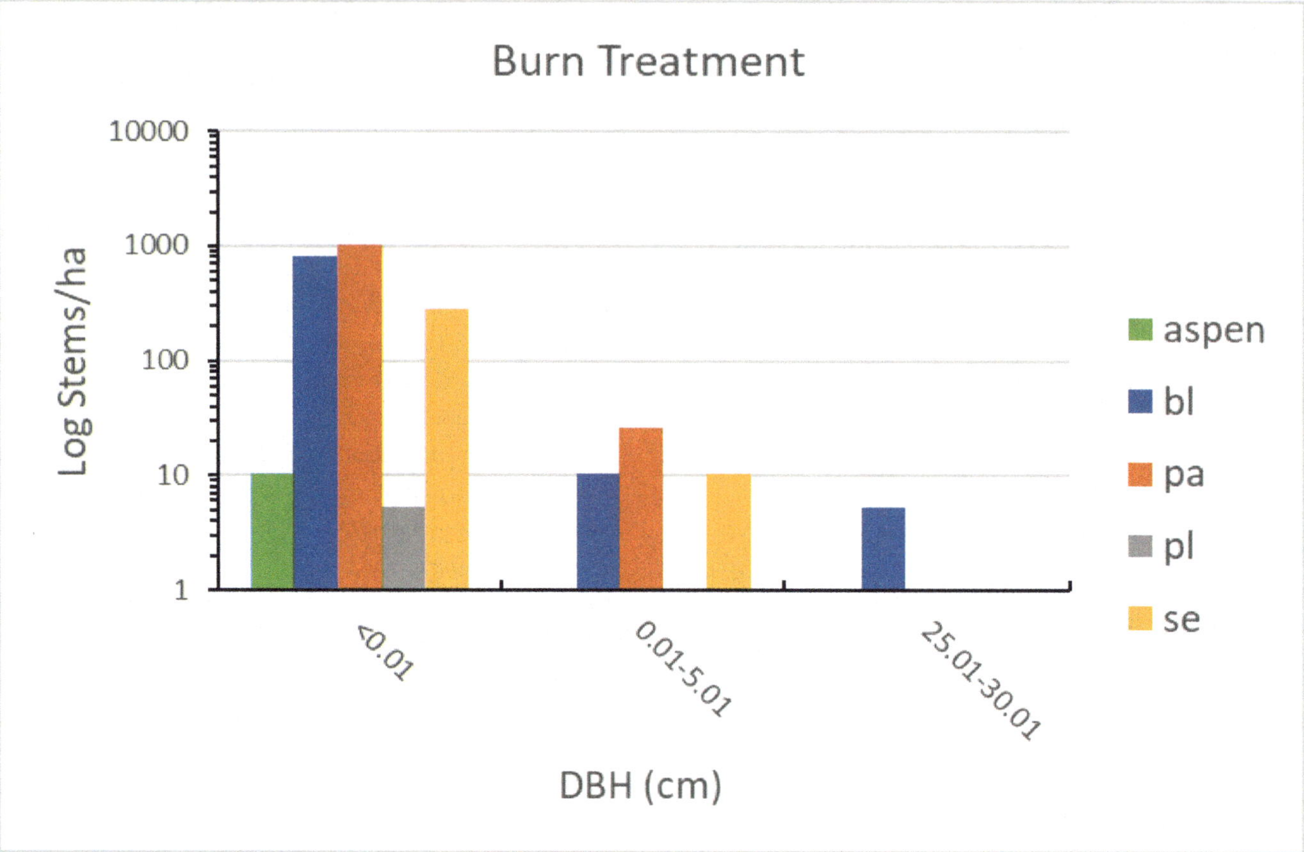

Figure 2. Helen Ridge burn treatment stem frequency distribution for tree species, Banff National Park, AB, Canada. bl = subalpine fir, pa = whitebark pine, pl = lodgepole pine, se = Engelmann spruce.

berry (figure 4). This mixture provided for significantly greater low shrub cover in these control plots, however, the strong abundance of grouseberry in these plots appeared to make up that difference in the dwarf shrub/herb layer.

MANAGEMENT IMPLICATIONS

Based on this 20-year remeasurement we conclude that this first Canadian whitebark prescribed burn was very successful in generating an increase in young healthy whitebark trees well above the levels found in adjacent undisturbed forest. We suggest that other efforts to reintroduce fire disturbance into these habitats for whitebark restoration continue with planned monitoring

The early successional floristic patterns indicate that the burn treatment was also less diverse, favouring *Vaccinium spp.* However, field observations indicate that there may be some small-scale association between a combination of understory species, micro-topography and regenerating whitebark vigor. Further analysis of these data will be explored.

ACKNOWLEDGEMENTS

This field work was made possible through funding provided by Parks Canada and the Natural Sciences and Engineering Research Council of Canada. Special thanks to Allison Fisher for field and technical support. Thank-you to Dr. Garth Mowat and Adrian Leslie for providing review comments on this paper.

LITERATURE CITED

B.C. Ministry of Forests and Range, BC Ministry of Environment, editors. 2010. Field manual for describing terrestrial ecosystems. 2nd ed. Victoria: Forest Science Program (Land management handbook).

Smith CM, B Wilson, S Rasheed, RC Walker, T Carolin, and B Shepherd. 2008. Whitebark pine and white pine blister rust in the Rocky Mountains of Canada and northern Montana. Canadian Journal of Forest Research. 38(5):982–995. doi:10.1139/X07-182.

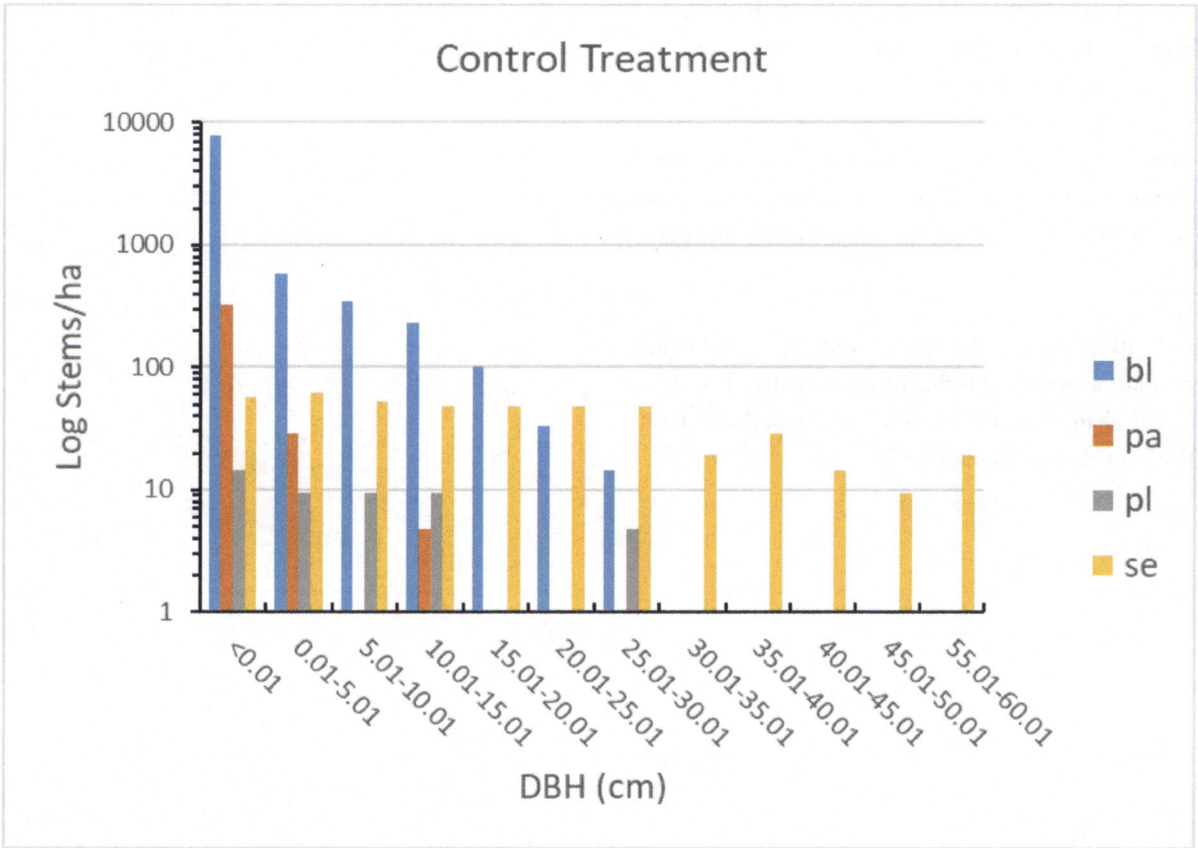

Figure 3. Helen Ridge control treatment stem frequency distribution for tree species, Banff National Park, AB, Canada. bl = subalpine fir, pa = whitebark pine, pl = lodgepole pine, se = Engelmann spruce.

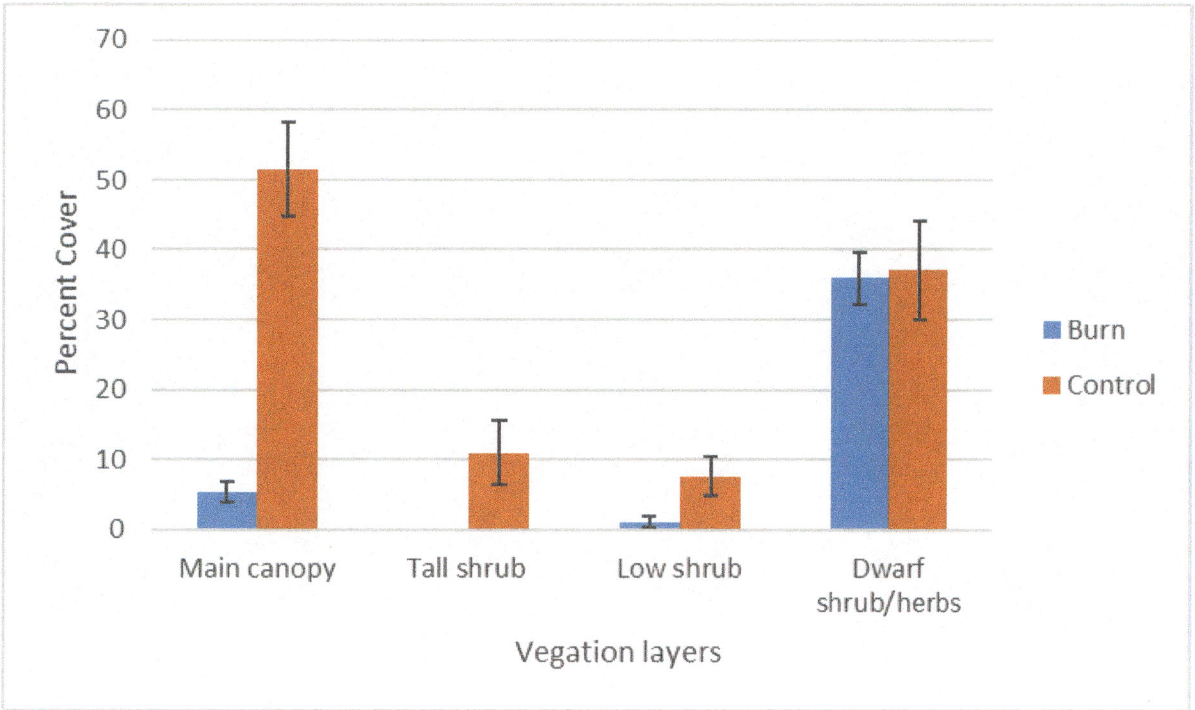

Figure 4. Vegetation cover values for Helen Ridge treatment units. Error bars are 95% confidence intervals.

Stuart-Smith, JG. 1998. Conservation of whitebark pine in the Canadian Rockies: blister rust and population genetics. Edmonton, AB: University of Alberta.

Wilson B, and J Stuart-Smith. 2002. Whitebark pine conservation for the Canadian Rocky Mountain national parks. Winlaw, BC: Cordilleran Ecological Research Report No.: KNP01- 01.

Wilson B, RC Walker, and J Stuart-Smith. 2002. Whitebark pine restoration and monitoring in the Canadian Rockies. In: Ecological and Earth Sciences in Mountain Areas. Banff, AB: The Banff Centre. p. 260–267.

Conference Abtracts

Differential Spring Budburst Phenology Across Western Five-Needled Pine Species

by Franklin Alongi | Danielle Ulrich | Department of Plant Sciences and Plant Pathology, Montana State University | Department of Ecology, Montana State University

Keywords: Great Basin Bristlecone pine, Limber pine, Whitebark pine, budburst, greenhouse, phenology

With conservation efforts of high-elevation pine species underway, understanding the mechanisms of spring growth initiation could optimize restoration efforts. We tracked the progression of seedling spring budburst in greenhouse-grown whitebark, limber, and Great Basin bristlecone pine seedlings of varying ages and from different source populations. We identified six stages of budburst phenology and assessed each seedling weekly starting from the end of February 2021 to present. Although the stages of spring budburst are still progressing, preliminary data show intriguing species, population, and age-level patterns. Limber pine advanced through budburst stages the fastest, followed by whitebark, with GB bristlecone pine advancing the slowest among the eldest age class. Limber pine exhibited strong within-species variation with one population largely responsible for the quick budburst progression. Interestingly, younger populations of whitebark pine show both the earliest and most rapid spring development progression. This study will improve restoration efforts and seedling establishment through improved selection of seedling populations to outplant based on budburst phenology. These improved selection criteria could help optimize the growing seasons and ultimate fitness of outplanted populations.

Soil Moisture Regime and Canopy Closure Structure Subalpine Understory Development Over 30 Years Following Stand-Replacing Fire

by Andrew J. Andrade | Diana F. Tomback | Timothy R. Seastedt | Sabine Mellmann-Brown | University of Colorado Denver | University of Colorado Denver | University of Colorado Boulder | United States Forest Service

Keywords: 1988 Yellowstone fires, succession, understory community

Western subalpine forests, critical habitat for many species of five-needle white pines, have experienced some of the most severe wildfires in recent decades due to climate change. Of particular concern is the potential for increasing soil moisture deficits to alter successional dynamics in post-fire communities. Recent studies indicate that conifer regeneration may be limited in some burns, but the cascading impacts on understory (forbs, graminoids, shrubs) community composition and microclimate remain unexplored. We investigated the long-term (30 years) associations among understory plant succession, soil moisture regime, and conifer regeneration in two study areas that burned during the 1988 Yellowstone fires. Permanent plots (n=275) were established in mixed whitebark pine (*Pinus albicaulis*) stands and classified by site type (mesic-burned, mesic-unburned, xeric-burned, xeric-unburned). Over the first decade of succession, species richness was lower in xeric-burned than in mesic-burned plots across study areas; however, conifer regeneration density was markedly higher at one study area. Nearly 30 years after fire at the study area where regeneration density was lower, soil temperatures and photosynthetically active radiation remained higher in burned than in unburned communities. There, understory community composition diverged by soil moisture regime, with the odds of graminoid cover 23-fold higher, and the odds of forb cover 3-fold lower, in xeric-burned than in mesic-burned plots. In contrast, at the study area where regeneration density was higher, microclimatic conditions converged between burned and unburned communities. Additionally, understory community composition did not diverge by differences in soil moisture regime but became more similar to unburned communities. Our results suggest that soil moisture regime structures the early trajectory of post-fire understory recovery, but the relationship diminishes as tree canopy closure alters microclimatic conditions. Under a warming climate, sparse tree canopy development may compound increasing aridity, with the potential for altered successional development of the understory.

Restoration Planting Options for Limber Pines (*Pinus flexilis* James) – 10 Years Later

by Anne Marie Aramati Casper | Kelly S. Burns | Anna W. Schoettle | Shannon Kay | Mountain Studies Institute & Colorado State University | USDA Forest Service, Forest Health Protection, Lakewood, CO | USDA Forest Service Rocky Mountain Research Station, Fort Collins, CO 80526 | USDA Forest Service Rocky Mountain Research Station, Fort Collins, CO 80526

Keywords: Cronartium ribicola, Limber pine, Pinus flexilis, artificial regeneration, bark beetles, facilitation,five-needle pines, planting, restoration

In 2018, 58.5% of seedlings were healthy, 7% higher than in 2012, as some seedlings with lower health ratings recovered. There continued to be significant differences in survival by planting site (75.2% to 30.3% living; X_2=274.96, df=5, p<0.0001). For the establishment period ('09-'12), seedlings performed best on the west and north sides of an object, and under denser canopy cover. For the growth period ('12-'18), seedlings performed best with a nurse object, regardless of orientation or percent canopy cover. These findings will be used to refine planting guidelines needed for management, conservation, and restoration efforts, including deploying WPBR-resistant seedlings. Importantly, our results indicate that unlike *P. albicaulis*, *P. flexilis* seedlings can be planted in at-risk stands, prior to overstory tree mortality. Therefore, planting methods are not universal for five-needle pine species.

My Introduction to Whitebark Pine

by Steve Arno | USFS, retired

Keywords: history, whitebark pine

I first saw whitebark pine while hiking in the Northeastern Olympic Mountains in 1960. By 1965 I had seen it in many more places, including the Washington Cascades, Oregon's Wallowa Mountains, the Sierra Nevada, and the Northern Rockies. Except in California and Oregon, it was already besieged by the introduced disease, white pine blister rust, and to some extent by mountain pine beetles. In 1970, while working as a forester on Northern Idaho's Kaniksu National Forest, forest ecologist Earle

Layser and I commiserated about whitebark's fate. Then while working on Bob Pfister's Habitat project, and later at Missoula's Fire Sciences Lab, I realized that whitebark is a fire-dependent species, unable to compete with other conifers, without periodic fires. At this time other forest fire ecologists had reached similar conclusions, but whitebark' cone crops were being recognized as a critical food source for the Clarks Nutcracker, squirrels and both black and grizzly bears. By the 1990s, biologists Diana Tomback, Kate Kendall, and ecologists Tad Weaver, Ward Maughey, and Ron Lanner, and young fire ecologist Bob Keane saw the need to form a Whitebark Pine Ecological Foundation (WPEF), to promote research and on-the ground activities, such as planting rust- resistant seedlings, being grown at the Coeur d' Alene by Mary Frances Maholovich. Under the leadership of Tomback, Keane and WPEF Canada, the organization has piled up accomplishments in recent years.

Managing High-Elevation Pines at Great Basin National Park

by Gretchen M. Baker | Great Basin National Park

Keywords: bristlecone, management, national park, threats

Great Basin National Park, located in east-central Nevada, contains over 10,000 acres of limber pine-bristlecone pine woodlands. Tens of thousands of visitors come each year to see ancient bristlecone pines (*Pinus longaeva*). Part of a visitor center exhibit is dedicated to them, as well as an interpretive walk through an ancient bristlecone grove. Current threats to bristlecone and limber pines are climate change, wildfire, mountain pine beetles, and the eventual arrival of white pine blister rust. The Park partners with the National Park Service Inventory and Monitoring Program to conduct surveys to characterize and study plots for long-term changes. In addition, the Park works with Anna Schoettle's Proactive Strategy. Seeds from bristlecone and limber pines are being tested for their resistance to white pine blister rust so that future trees can be grown from resistant trees.

Defense Characteristics Among High-Elevation Pines and Vulnerability to Native Bark Beetles

by Barbara Bentz | Rocky Mountain Research Station

Keywords: bristlecone pine, climate, mountain pine beetle

Changing climate has increased awareness of the native bark beetle mountain pine beetle (*Dendroctonus ponderosae*) in high-elevation five needle pine ecosystems. True to its name, mountain pine beetle has adapted to high-elevation mountain climates, with the capacity for population growth during warm periods. Warming in recent decades at high-elevations has facilitated mountain pine beetle population success and associated pine mortality in some forests. Although the majority of pine species are susceptible to mountain pine beetle attack and associated tree death, vulnerability varies among high-elevation pine species. We describe and compare defense capacity and attack vulnerability among Great Basin bristlecone (*Pinus longaeva*), foxtail (*P. balfouriana*), Rocky Mountain bristlecone (*P. aristata*) and limber (*P. flexilis*) pines. We also evaluate causal agents of recent mortality in Great Basin bristlecone pine, a species considered the least susceptible to mountain pine beetle attack and reproduction. Understanding the role of native insects and climate in mortality of iconic high-elevation pines is vital for protecting and projecting the future of these foundational ecosystems.

Divergent, Age-Associated Fungal Communities of *Pinus Flexilis* And *Pinus Longaeva*

by Joseph D. Birch | James A. Lutz | Benjamin L. Turner | Justine Karst | Department of Renewable Resources, University of Alberta, Edmonton | Department of Wildland Resources, Utah State University, Logan, Utah | Soil and Water Science Department, University of Florida, Gainesville, FL, United States | Department of Renewable Resources, University of Alberta, Edmonton

Keywords: Great Basin bristlecone pine, Limber pine, belowground ecology, dendroecology, ectomycorrhizal, facilitation, symbiotic

The long-lived five-needle pines, *Pinus flexilis* (limber pine) and *Pinus longaeva* (Great Basin bristlecone pine) can co-occur and may form symbiotic partnerships with the same species of ectomycorrhizal fungi. These shared symbiotic relationships may facilitate the persistence of these pine species. Throughout their lives, *P. flexilis* and *P. longaeva* may also assemble unique belowground fungal communities, adding to the conservation value of ancient trees. We used MiSeq sequencing of fungal rDNA to compare fungal community similarity for co-occurring *P. flexilis* and *P. longaeva* roots and soils in an old-growth forest at the Utah Forest Dynamics Plot, Utah, USA. We cored trees to measure their age and determine whether fungal communities change with advanced tree age. We found 720 amplicon sequence variants associated with *P. flexilis* roots, 736 with P. longaeva roots, and 199 that were shared between the two pines. Root-associated fungal communities were significantly different between *P. flexilis* and *P. longaeva* despite similar soil communities. The fungal community composition on *P. flexilis* roots and around *P. longaeva* soil was associated with advanced tree age up to 1340 years. The root-associated fungal community of *P. flexilis* and the soil community of *P. longaeva* increased in dissimilarity with tree age, indicating that age heterogeneity within old-growth stands promotes fungal diversity. The significant differences in root-associated fungal communities between the two pine species highlights that they are likely engaged in different bi-directional selection with fungal communities.

Limber Pine Restoration in the Black Hills National Forest, South Dakota

by James T Blodgett | Cheryl Mayer | USDA Forest Service | USDA Forest Service

Keywords: branch pruning, Cronartium ribicola, limber pine, management, monitoring, planting, restoration, seed collection, white pine blister rust

As part of the Limber Pine Restoration Project, a new limber pine population was established in 2017 in the Norbeck Wildlife Preserve, Black Hills National Forest. In South Dakota, limber pine (*Pinus flexilis* E James), a Black Hills National Forest species of local concern, occurs in isolated areas scattered over a small geographic area of about 518 ha in the Black Elk Wilderness of the forest and adjacent Custer State Park. Recently many of these pines were killed by mountain pine beetle and white pine blister rust. An integrated management plan has been implemented in the forest including planting two-year-old limber pine seedlings. Four hundred and fifty-five seedlings from local seed were planted in the spring of 2017, 2018, and 2021 at seven areas with the help of several volunteers. Vexar tubes (i.e., animal protectors) were staked around seedlings to protect them from herbivory. Seedling survival was better than expected (97%) and we are seeing good exponential growth (average 6.1 cm/yr). Most of the seedlings are growing well and the Vexar tubes appear to be protecting seedlings. These sites will be monitored over the next few years.

Mechanisms of Vegetation Change in High-Elevation Forests of the Greater Yellowstone Ecosystem

by Erika Blomdahl | James H. Speer | Margot Kaye | Nicole E. Zampieri | Maegen Rochner | Bryce Currey | Denise Alving | Cahalan | Ben Hagedorn | Hang Li | Rose Oelkers | Lissa Pelletier | Ichchha Thapa | Kevin Willson | Brian D. Woodward | R. Justin DeRose | Utah State University | Indiana State University | Penn State University | Florida State University | University of Louisville | Montana State University | Penn State University | The Nature Conservancy | Western Washington University | Indiana State University | Lamont-Doherty Earth Observatory | SUNY-ESF | Indiana State University | University of New Mexico | Colorado State University | Utah State University

Keywords: Greater Yellowstone Ecosystem, dendrochronology, drought, ecotone shift, global change, high-elevation, whitebark pine

Global climate change is predicted to cause widespread changes in the distribution of forest vegetation, particularly in mountain environments where climate exerts strong controls on community arrangement. The upslope movement of vegetation communities has been observed in association with warming temperatures, especially evident in ecotones, or transition zones between vegetation types. We explored the role of drought and tree mortality on recent change in high-elevation forests of the Greater Yellowstone Ecosystem (GYE). We established 20 forest demography plots along an elevational gradient spanning the dominant high-elevation vegetation types in the GYE. Establishment dates indicated ecotone shift from meadow to forest, where *Pinus albicaulis* and *Pinus contorta* moved into the highest elevations following the 1950s. Among forest types, comparisons between live overstory and understory compositions suggested an upslope movement of *Abies lasiocarpa*, stability in *Pinus contorta*, and counterintuitively, downslope movement of *Pinus albicaulis*. Tree mortality was concentrated in the *Pinus* spp., largely due to *Dendroctonus ponderosae* activity from 2008-2012 that was exacerbated by drought conditions. Interestingly, the primary driver of growth variability in both *Pinus* spp. shifted from temperature in spring and winter to drought relatively rapidly during the 1950s. It appears that increased sensitivity to drought during the mid-20th century, in combination with increasing stand density associated with aging forests, created conditions of increased susceptibility to drought-related beetle-caused mortality during the most recent drought, the largest in the last millennia. While the species in these high-elevation forests of the GYE responded individually to expected successional processes, global change stressors (i.e., drought) acted on these forests in a complex way, and provided evidence for both ecotone shifts and stability.

Conservation and Restoration of Whitebark Pine by the USFS in Oregon and Washington

by Andrew Bower | USDA Forest Service

Keywords: USFS Pacific Northwest Region, conservation, restoration, restoration strategy, rust resistance

There are over 1.1 million acres of whitebark pine habitat in Oregon and Washington, over 90% of which is on U.S. Forest Service land. The US Forest Service Pacific Northwest Region (OR and WA) has allocated substantial financial and human resources in efforts for conservation and restoration of whitebark pine. These efforts have followed a plan outlined in a document titled the "Whitebark Pine Restoration Strategy for the Pacific Northwest Region" which presents a comprehensive plan to reach the goal of "a network of viable populations of whitebark pine throughout the Pacific Northwest". The key actions prescribed include:

- collect seed for rust resistance screening, reforestation, and gene conservation refine species habitat mapping and assess stand conditions
- planting seedlings in areas where natural regeneration is lacking
- thin or remove competing vegetation to release whitebark pine and reduce fuel loads treat for mountain pine beetle in areas experiencing outbreaks
- continue a rust screening program

Highlights of some of the activities that have been undertaken as prescribed in this strategy as well as future direction, including how the Pacific Northwest Regional Restoration Strategy will tie in with the National Whitebark Pine Restoration Plan will be discussed.

Does High Severity Fire Facilitate Species Transition in California Subalpine Forests?

by Emily Brodie | Hugh Safford | Joseph Stewart | Andrew Latimer | Jesse Miller | UC Davis, Department of Environmental Science and Policy | USDA Forest Service | UC Davis, Department of Environmental Science and Policy | UC Davis, Department of Plant Sciences | Stanford, Biology Department

Keywords: Fire severity, species transition, subalpine forest, tree regeneration

Many montane tree species distributions are likely to move upslope in the coming decades due to predicted changes in precipitation and temperature. However, long-lived species such as trees might lag in colonizing climatically suitable habitat because adult trees can persist for long periods in stressful environments. By removing adult trees and reducing the "biological inertia" of an intact forest, disturbances such as fire are predicted to facilitate changes in species composition. In this study we asked whether high severity fire facilitates regeneration of non-dominant lower elevation tree species in subalpine forest in California. California subalpine forest is of particular interest because climate-driven changes in stand density and annual burned area are already occurring in this forest type. Preliminary analysis of tree regeneration in 248 plots across 13 fires suggests that high severity fire does not facilitate regeneration of lower elevation species over higher elevation species. In fact, while white pine regeneration generally increased with fire severity, lower elevation species (predominantly red fir) regeneration decreased with fire severity. In the Sierra Nevada, where shade-tolerant red firs dominate just downslope from subalpine forest, the existence of appropriate regeneration microsites seems to be having more of an effect on species regeneration patterns than broad-scale shifts in temperature. While red fir may be gaining a foothold in shadier, unburned subalpine forest understories, white pines do better in exposed post-fire sites with high canopy mortality. Our results suggest that small patches of high severity fire such as those sampled in this study might help to maintain white pine stands in the face of upward movement of lower elevation species that are shade tolerant.

Can We Improve Western White Pine Microbiomes to Promote Resistance to Blister Rust Disease?

by Lorinda Bullington | Emily Martin | Nadir Erbilgin | Peter Kennedy | Richard Sniezko | Department of Ecosystem and Conservation Sciences, University of Montana, Missoula, MT, USA | MPG Ranch, Missoula, MT 59801, USA 598801, USA | Department of Renewable Resources, University of Alberta, Edmonton. Alberta, Canada | Department of Plant and Microbial Biology, University of Minnesota, St. Paul, MN, USA | USDA Forest Service, Dorena Genetic Resource Center, Cottage Grove, OR 97424, USA

Keywords: Cronartium ribicola, Endophytes, ectomycorrhizal fungi, terpenes

Hundreds of asymptomatic fungal taxa live inside healthy white pine tissues. Recent studies suggest that these fungi can influence the frequency and severity of infections by fungal pathogens such as *Cronartium ribicola*, the causal agent of white pine blister rust. In a full- factorial experiment, we inoculated western white pine (*Pinus monticola*) seedlings from six seed families with foliar fungal endophytes (FFE) or ectomycorrhizal fungi (EMF, genus: *Suillus*), as well as a combined fungal treatment (FFE+EMF), and a control treatment (no inoculation). The six seed families consisted of half-siblings and full-sib progeny, with both high and low levels of known disease resistance. Four months post-inoculations, we infected a subset of seedlings from all four treatments with the rust to determine the individual and shared effects of FFE and EMF on resistance to white pine blister rust as well as seedling performance. We measured tree defensive chemistry (terpenes) for all treatments immediately before, and four months after infection, as disease symptoms began to appear. Seed family influenced both initial disease severity and terpene composition in all seedlings. We observed a strong treatment effect on tree defensive terpenes after pathogen infection ($p < 0.005$), indicating that fungal inoculations altered seedlings' induced defense responses. EMF inoculations had the greatest influence, while EMF+FFE treated seedlings were more similar to untreated control seedlings, suggesting an interaction between above and belowground fungi. How these differences in induced defensive responses translate to disease resistance will be assessed with ongoing monitoring of disease progression.

Efficacy of Pruning Limber Pine to Mitigate White Pine Blister Rust Impacts

by Kelly Burns | USDA Forest Service, R2-FHP

Keywords: disease management, limber pine, pruning, white pine blister rust

White pine blister rust is a disease that damages and kills white pines in the Southern Rocky Mountains. The purpose of this study was to determine if various pruning treatments could prevent crown dieback and increase the longevity of high-value limber pines in two recreation areas with high disease pressure. Four treatments were applied to infected trees:
(1) do nothing (control); (2) preventive pruning of branches and needles up to 7 ft (2.1 m) or 60% of the crown, whichever was less; (3) preventive and sanitation pruning (removing cankers throughout the crown); or (4) sanitation pruning only. Treatments for uninfected trees included control or preventive pruning. Trees were assessed and treated three times over the course of the study (2005-2006, 2010-2011, 2015). The same treatments were reapplied if new cankers were detected during subsequent assessments. Data collected included tree health status; branch and stem canker counts; type (stem or branch), number, and crown/stem location for canker removals; and branch canker lengths (to infer infections periodicity). Percent crown dieback

was quantified in the last two assessments. Significantly less crown dieback occurred on trees that received both preventive and sanitation pruning than on controls. Sanitation pruning removed the most cankers. New cankers occurred annually on 47–52% of originally infected trees and 10% of originally uninfected trees, suggesting that a 5-year pruning rotation is needed on sites with high disease pressure. A longer rotation could be used in areas with lower disease pressure. Management guidelines are provided.

Verbenone and Green-Leaf Volatiles Reduce Whitebark Pine Mortality in a Northern Range-Expanding Mountain Pine Beetle Outbreak

by Etienne Cardinal | Brenda Shepherd | Jodie Krakowski | Parks Canada | Parks Canada | Independent Consultant

Keywords: mountain pine beetle, verbenone, whitebark

This is the first study testing effectiveness of pheromone treatments to protect individual trees from a range-expanding mountain pine beetle (MPB, *Dendroctonus ponderosae* Hopkins) attack into newly exposed host populations of endangered whitebark pine (*Pinus albicaulis* Engelmann). We investigated the effectiveness of a combination of verbenone and Green-Leaf Volatiles (GLV) to protect rare and valuable disease-resistant trees during a MPB epidemic from 2015 to 2018 in Jasper National Park, Canada. Treatments reduced the proportion of trees attacked by MPB for all diameter classes, across all stands, from 46 to 60%. We also evaluated the effect of the exotic disease white pine blister rust, the species' other main regional threat. MPB were less likely to attack large, infected trees than healthy trees, emphasizing the value of the pheromone treatment. Protecting large, cone-bearing disease-resistant whitebark pine trees is fundamental to whitebark pine recovery. Maintaining reproductive trees on the landscape increases the frequency and diversity of rust-resistant genotypes more effectively than just planting seedlings to replace MPB-killed trees, because this slow-growing species takes over 80 years to reproduce. Our study confirmed protecting large rust-resistant trees with verbenone and GLV is a proactive and effective treatment against MPB for whitebark pine in naïve populations.

Early Results from Whitebark Pine Genecology Field Trials in British Columbia

by Charles V Cartwright | Richard A Sniezko | Michael P Murray | Iain R Reid | BC Ministry of Forests | USDA Forest Service | BC Ministry of Forests | Parks Canada

Keywords: genecology, genotype by environment interaction, whitebark pine

Whitebark pine (*Pinus albicaulus* Englm.) is listed as endangered in Canada, chiefly due to decimation by white pine blister rust (*Cronartium ribicola* J. C. Fisher), and mountain pine beetle (Drendroctonus ponderosae Hopkins). In response to the federal listing of the species, it is incumbent on the Province of British Columbia (BC) to take steps towards restoration. Multi-environment genecology field trials were established with the intent of screening for blister rust resistance trees, quantifying genotype by environment interactions, providing a basis for developing climate-based seed transfer guidelines, and eventually evidence concerning trait durability. Test sites were established in most of the climate types where whitebark grows in BC using seed sources from across much of the range. The same provenances were also deployed in a farm field setting and some were scored in controlled inoculation screening. Effects of seed source latitude, elevation, and several climate variables on survival and growth are reported. As well, correlation of rust resistance scores across some of the sites is described. Implications for restoration using rust resistant planting stock in BC are considered.

Predicted Impacts of Climate Change on the Northern Range Limit of Whitebark Pine

by Alana J Clason | Eliot J.B. McIntire | Philip J. Burton | Bulkley Valley Research Centre | Natural Resources Canada | University of Northern BC

Keywords: Clark's nutcrackers, Climate change, dispersal, latitudinal range limit, suitable habitat

A common expectation for tree adaptation to climate change involves elevational or latitudinal migration to track suitable climate. Whitebark pine reaches its northern range limit in north-central BC, with many bioclimatic species distribution models predicting currently suitable climate further north than this range limit. We tested the role of climate across scales on the occurrence and abundance of whitebark pine at its northern limit and found it is commonly associated with cold habitats throughout this northern range. These results suggest cold is not limiting the species, and migration north with climate change is unlikely with increasing temperatures alone. The biotic interaction between whitebark pine and Clark's nutcrackers, and between the nutcracker and its alternative food sources such as Douglas-fir, drive current northern range dynamics, and these interactions will also determine range shifts with climate change. While cold temperatures are not currently limiting, climate change will reduce suitable habitat throughout the range of whitebark pine, with much of future suitable climate habitats occurring further north than the current range. Blister rust, mountain pine beetle and fire all result in whitebark pine mortality throughout its northern range, suggesting restoration of these northern ecosystems to sustain the pine-nutcracker interaction will be required to facilitate future dispersal north.

Clark's Nutcrackers Influence on the Northern Range of Whitebark Pine

by Alana J. Clason | Eliot J.B. McIntire | Philip J. Burton | Bulkley Valley Research Centre | Natural Resources Canada | University of Northern BC

Keywords: Clark's nutcrackers, Northern range limit, biotic interactions, dispersal, disturbance, whitebark pine distribution

The co-evolved positive biotic interaction between whitebark pine and Clark's nutcrackers influences the occurrence and abundance of whitebark pine across scales. The pine- nutcracker interaction may drive the current northern limit of whitebark pine in north- central BC, influenced by the distribution of alternative food sources (Douglas-fir). Whitebark pine is suffering ongoing mortality throughout its northern range from blister rust, mountain pine beetle, and fire. On top of these multiple stressors, maintaining habitat for Clark's nutcrackers will be required to sustain whitebark pine dispersal within this range and to facilitate migration into habitats that will be less vulnerable to climate change. Here, we will present evidence for the role of nutcrackers in limiting the northern range of whitebark pine. We also suggest potential impacts to future whitebark pine distributions across the northern range through cumulative disturbances that could disrupt pine- nutcracker interactions.

Limber Pine Condition on Montana's Rocky Mountain Front

by Christy M. Cleaver | Katie M. McKeever| Amy M. Gannon | August C. Kramer | Dave Hanna | USDA Forest Service, Forest Health Protection | USDA Forest Service, Forest Health Protection | Montana Department of Natural Resources | Montana Department of Natural Resources | The Nature Conservancy

Keywords: limber pine, white pine blister rust, mountain pine beetle, Montana

White pine blister rust (WPBR) has caused extensive crown dieback and mortality along Montana's Rocky Mountain Front since its introduction to eastside limber pine in the 1930's. The combined impacts of WPBR, mountain pine beetle, and changing climate patterns are suspected to be contributing to mortality and the alteration of limber pine stands occupying the grassland-montane ecotone throughout this region. Information on stand conditions is needed to inform management and restoration efforts. Our study objectives were to: (1) assess site and stand characteristics that describe limber pine along Montana's Rocky Mountain Front (2) determine the status and health of limber pine trees and regeneration, and (3) characterize the major damage agents on limber pine trees and regeneration and determine the occurrence, incidence, and severity of WPBR on limber pine.

In 2017 and 2018, we assessed 4,427 limber pine on 74 plots in limber pine-dominated stands along a 50-mile latitudinal gradient west of Choteau, Montana. Mean density of live limber pine was 150 trees per acre. Thirty-seven percent of limber pines were classified as healthy, 21% were declining or dying, and 43% were dead. White pine blister rust was the primary damage agent, occurring in 100% of plots with a mean incidence of 36%. Bark beetle-caused mortality was low at less than 5%. Live limber pine seedling density averaged 251 trees per acre. Of all limber pine seedlings 69% were alive, 31% were dead, and WPBR occurred on 5% of live seedlings. Competing Douglas-fir trees and seedlings were common. This establishment of baseline conditions for Montana's Rocky Mountain Front limber pine stands will help to monitor effects of future climate change, assist resource managers in forecasting WBPR impacts in areas more recently invaded by the disease, and inform future restoration activities for limber pine.

Evaluating Spectroscopy as a Tool For Rapid Identification of Wpbr-Resistant Whitebark Pine Trees

by Anna O. Conrad | Caterina Villari | Richard Sniezko | Pierluigi Bonello | USDA Forest Service | University of Georgia | USDA Forest Service | The Ohio State University

Keywords: Pinus albicaulis, machine learning, predictive models, rapid phenotyping, resistance chemotypes, support vector machine, white pine blister rust

One of the major challenges in selecting and breeding for white pine blister rust (WPBR) - resistant whitebark pine trees lies in the initial phenotyping phase. Currently available methods require time consuming and resource intensive inoculations of progeny, followed by months or years of observation to determine which individuals are resistant and which susceptible. Here we show that predictive models based on machine learning (ML, AKA artificial intelligence) applied to Fourier-transform infrared (FT-IR) and Raman spectra of plant tissues, or their extracts, can be useful to streamline and greatly accelerate this all- important initial phase of the process. The best ML model we produced, based on support vector machine (SVM), correctly classified 76.7% of the FT-IR spectra as coming from either resistant or susceptible trees. With Raman spectra, the accuracy of the best SVM model was 67.9%. Both accuracies are significantly better than chance alone.

In conclusion, SVM appears to hold great promise as a rapid predictive tool for WPBR resistance when combined with FT-IR or Raman spectroscopy. Our results are already encouraging, but the inclusion of spectra from additional WBP with known ratings would likely strengthen the predictive power of ML models even further, and allow for additional validation and refinement, ultimately leading to a more robust and reliable, yet fast and non- destructive, predictive tool. Further investments in these areas of phenotyping technology, as well as in the very promising area of near infrared (NIR) spectroscopy, could result in field deployable, user-friendly hand-held sensors in the near future, and are strongly encouraged.

Post-Fire Regeneration of Endangered Limber Pine (*Pinus Flexilis*) at the Northern Extent of Its Range

by Denyse A. Dawe | Vernon S. Peters | Mike D. Flannigan | Natural Resources Canada | King's University| University of Alberta

Keywords: Limber pine; Pinus flexilis, fire ecology, post-fire regeneration

Limber pine (*Pinus flexilis*), an understudied tree species important to montane and subalpine ecosystems, is listed as endangered in Alberta. Prescribed burning is a proposed tool to stimulate natural regeneration via dispersal by Clark's nutcracker (*Nucifraga columbiana*). Extensive burns are suggested as a way to provide limber pine with a competitive advantage, due to long-distance dispersal. However, no studies had surveyed limber pine's immediate post-fire regenerative response at the northernmost extent of its range. We examined post-fire regeneration in one 16-year-old prescribed burn and one 8- year-old wildfire at the northern edge of limber pine's distribution. We then compared this to a regeneration baseline of plots in nearby unburned limber pine acting as seed sources to the burns.

Overall, we found only six post-fire limber pine seedlings within the burns compared to one hundred twenty-four similarly aged seedlings found in unburned plots. Statistical modelling suggests that distance to a seed source and microclimate effects, due to lack of canopy cover, may have influenced these low seedling numbers in our burned plots. Our findings suggest that extensive prescribed burns and stand-replacing wildfire events at the northern extent of limber pine's range may have limited seedling regeneration in the immediate post- fire period. Conservation efforts may be well-served by focusing on fire mitigation practices, such as thinning and other fuel treatments, in areas surrounding established limber pine stands. Finally, post-disturbance stands could benefit from supplementary seedling plantings to achieve recovery plan restoration goals for this endangered species.

Variable Demographic Patterns Interact with Disturbance to Shape Limber Pine Population Viability

by R. Justin DeRose | Utah State University

Keywords: climate change adaptation, generalist, mountain pine beetle, nutcrackers, surrogate species

Limber pine is an important western tree species having the largest ecological amplitude of any of the five needle pines. Its presence in western forests is equally likely irrespective of elevation, aspect or slope. Recent, severe, and widespread decline of a closely related species, whitebark pine, is suggestive of a similar fate for limber pine. Indeed, recent work has indicated that range-wide mortality in limber pine is approaching similar levels as was seen in whitebark pine two decades ago. Mountain pine beetle outbreaks are likely responsible for the vast majority of limber pine mortality. Whether the recent mountain pine beetle outbreaks are novel, or a normal part of disturbance dynamics in limber pine forests is unknown. In this study, we examine the long-term persistence of a limber pine woodland located in the Greater Yellowstone Ecosystem. Dendroecological methods were employed to reconstruct establishment dates, death dates, and cause of death for 78 total (n=28 live, and n=50 dead) limber pines. Over the past ~800 years, limber pine recruitment was characterized as pulsed, with peaks occurring in the 13th and 17th century consistent with multi-decadal pluvial periods. Over the past ~700 years limber pine mortality was much more continuous, but with pulses centered on the 1880s, 1960s, and early 2000s. Death of over half the limber pine since ~1860 can be attributed to mountain pine beetle. Because continuous propagule pressure for limber pine is driven by nutcracker, the pulses of successful establishment we saw appear to have been driven by relatively wet periods. In contrast, background mortality rates were very low, and pulses of mortality driven by the mountain pine beetle. The balance between continuous and episodic demographic processes in the face of mounting disturbance pressure over time is likely to further bottleneck limber pine population viability.

Special Session: The North American Dendroecological Fieldweek (NADEF): Educational and Research Opportunities to Better Understanding 5-Needle Pines

Organizer: Maegen Rochner

The Importance of Energy-Water Limitation Threshold in Drought Impact Studies

by Joan C. Dudney | University of California, Santa Barbara

Keywords: climate change, drought, tree-rings, whitebark pine

Forest diebacks have increased in magnitude in many regions in response to greater water limitation. "Hotter droughts" are predicted to increase under climate change and result in significant restructuring of forest composition and ecosystems services. Many drought- related studies, however, focus on water limited systems, where water—not energy—poses the greatest constraint on photosynthesis. Contrary to expectation, hotter, drier conditions in energy limited systems may cause greater growth, as the growing season is extended. Thus, identifying the location of the energy-water limitation threshold is critical to predict forest mortality under climate change. Here we assess the impacts of the recent extreme drought in California (~2012-2015) on subalpine whitebark pines across the central and southern Sierra Nevada. We use a combination of over 700 tree-rings and over 1,000 stable isotope samples to test whether extreme drought led to greater growth or greater physiological stress. We show that during extreme drought, the energy-water limitation threshold shifted upslope into higher elevation whitebark pine. Trees growing near this threshold experienced some physiological stress, but trees far from this threshold experienced positive growth. These results suggest that extreme drought has a more nuanced effect on average productivity for forests that occur across strong climatic gradients.

Nonlinear Shifts in White Pine Blister Rust Due to Climate Change

by Joan C. Dudney | John Battles | Adrian Das | Jonathan Nesmith | Claire Willing | Andrew Latimer | University of California, Santa Barbara | UC Berkeley | USGS | US Forest Service | Stanford University | UC Davis

Keywords: White pine blister rust, climate change, foxtail pine, range expansions, sugar pine, western white, whitebark pine

Though climate change is predicted to cause major shifts in infectious disease risk, definitive evidence is often elusive due to data limitations and confounding factors. Thus, disease outbreaks are often interpreted as stochastic events, rather than a response to changing environmental conditions. Nonlinearities in climate change patterns can also complicate inference of mechanistic drivers. Nonlinear stochastic events, such as droughts, are also predicted to increase in frequency and severity throughout various parts of the globe and negatively affect fungal pathogens. Here we take advantage of a unique long-term dataset (two survey periods spanning ~19 years; over 8,000 individual hosts) of the fungal tree disease, white pine blister rust (*Cronartium ribicola* Fisch., blister rust) in Sequoia and Kings Canyon National Parks. We find that climate change between 1996 and 2016 moved the climate optimum of the disease into higher elevations. The nonlinear climate change- disease relationship contributed to an estimated 5.5 (4.4-6.6) percentage points (p.p.) decline in disease prevalence in arid regions and an estimated 6.8 (5.8-7.9) p.p. increase in colder regions. Though climate change likely expanded the suitable area for blister rust by 777.9 (1.0-1392.9) km2 into previously inhospitable regions, the combination of host- pathogen and drought-disease interactions contributed to a substantial decrease (32.79%) in mean prevalence between surveys. Specifically, declining alternate host abundance suppressed infection probabilities at high-elevations, even as climatic conditions became more suitable. Further, drought-disease interactions varied in strength and direction across an aridity gradient—likely decreasing infection risk at low elevations while simultaneously increasing infection risk at high-elevations. These results highlight the critical role of aridity in modifying host-pathogen-drought interactions. Variation in aridity across topographic gradients can strongly mediate plant disease range shifts in response to climate change.

Compounding Effects of White Pine Blister Rust, Mountain Pine Beetle, and Fire on White Pines in Sequoia and Kings Canyon National Parks

by Joan Dudney | Jonathan Nesmith | Matt Cahill | Jennifer Cribbs | Dan Duriscoe | Adrian Das | Nathan Stephenson | John Battles | UC Davis | USFS PNWRS | The Nature Conservancy | UC Davis | NPS Night Sky Program | USGS WERC | USGS WERC | UC Berkeley

Keywords: I&M, NPS, Sierra Nevada, white pines

Invasive pathogens and bark beetles have caused precipitous declines of white pines in North America. We characterized long-term patterns of mountain pine beetle (*Dendroctonus ponderosae*; MPB) attacks and white pine blister rust (WPBR; *Cronartium ribicola*). We focused on four dominant white pine host species in Sequoia and Kings Canyon National Parks (SEKI), including sugar pine (*Pinus lambertiana*), western white pine (*P. monticola*), whitebark pine (*P. albicaulis*), and foxtail pine (*P. balfouriana*). Between 2013 and 2017, we resurveyed 152 long-term monitoring plots that were first established between 1995 and 1999. Overall extent (plots with at least one infected tree) of WPBR increased from 20% to 33%. However, the infection rate across all species decreased from 5.3% to 4.2%. WPBR dynamics varied greatly by species, as infection rate decreased from 19.1% to 6.4% in sugar pine, but increased in western white pine from 3.0% to 8.7%. For the first time, WPBR was recorded in whitebark pine in SEKI, but foxtail pine remained uninfected. MPB attacks were highest in sugar pines and decreased in the higher elevation white pine species, whitebark and foxtail pine. Both WPBR and MPB were important factors associated with elevated mortality in sugar pines. In addition, multiple mortality agents, including WPBR, MPB, fire, and drought contributed to major declines in sugar pine and western white pine; recruitment rates were much lower than mortality rates for both species. Our results highlighted that sugar pine has been declining much faster in SEKI than previously documented. Given current spread patterns, blister rust will likely continue to increase in higher elevations, threatening subalpine white pines in the southern Sierra Nevada. More frequent long-term monitoring efforts could inform ongoing restoration and policy focused on threats to these highly valuable and diverse white pines.

Special session title: White pines in NPS. Special session organizer: Kristin Legg

Climatic Factors Affecting White Pine Blister Rust Incidence on Whitebark Pine in Washington National Parks

by Sebastian Espinosa Novoa | Beth Fallon | John Boetsch | Northern Arizona University, Department of Earth and Sustainability; National Park Service, North Coast and Cascades Inventory and Monitoring | National Park Service, Mount Rainier National Park | National Park Service, North Coast and Cascades Inventory & Monitoring Network

Keywords: National Park Service, blister rust, climate drivers, whitebark pine

Whitebark pine, *Pinus albicaulis*, is a low-cover but critically important tree in the subalpine ecosystems of national parks within western Washington state, in the North Coast and Cascades Inventory & Monitoring Network (NCCN). Trees within North Cascades (NOCA) and Mount Rainier National Parks (MORA) have experienced mortality rates ranging from 9-45% and 21-50%, respectively. Mountain pine beetle attacks are rare and mortality is attributed primarily to white pine blister rust (WPBR). The interpark differences in mortality may be partially attributable to different climates between NOCA, at a higher latitude, and MORA; past analysis of monitoring data has demonstrated a correlation of increasing latitude with declining mortality (Rochefort et al. 2018). In the present study we explore how environmental factors are driving WPBR rates within the remaining living tree populations at MORA and NOCA and assess intrapark climatic differences that are correlated with particularly vulnerable populations and/or areas ripe for restoration work (e.g., refugia). As in other locations, seasonal variation in moisture and temperature may increase disease presence in whitebark pine. Analyses of NCCN monitoring plots show that while Park is a significant factor in mortality rates, vapor pressure deficits in August are another significant predictor ($P < 0.05$), but simple models of vapor pressure, location, temperature and precipitation only explain 32% of the variation in mortality. WPBR infection rates of live trees are not significantly predicted by the same models. Misidentification of blister rust, field crew turnover, and coarse environmental data may be reducing our ability to model environmental variables affecting rust in these parks. In 2021, we will complete establishment of new monitoring plots and catalog tree disease status in all existing plots, and b) model disease incidence with higher resolution climate data to evaluate the abiotic conditions contributing to current WPBR presence and decline in NCCN parks.

Comparative Species Assessments of Five-Needle Pines Throughout the Western United States

by Sara A. Goeking | Marcella A. Windmuller-Campione | US Forest Service | University of Minnesota

Keywords: Great Basin bristlecone, Pinus albicaulis, Pinus aristata, Pinus balfouriana, Pinus flexilis, Pinus longaeva, Pinus strobiformis, Rocky Mountain bristlecone pine, forest inventory, foxtail, limber, southwestern white, whitebark

Five-needle white pine species provide important ecosystem services throughout the western U.S., and many of these species have become susceptible to stressors including warmer temperatures, insect epidemics, nonnative disease, and altered disturbance regimes. The objective of this study was to characterize recent broad-scale demographic patterns, including species abundance (i.e., numbers of individuals, tree density, size-class distributions, recruitment, growth rates, mortality rates, and causes of mortality, for the six species of five-needle pine that occur in the western US. We used the U.S. Forest Service's Forest Inventory and Analysis (FIA) dataset, spanning >10 years, to quantify demographic status and trends for each species. FIA data were compiled from a probabilistic sample design and consistent analysis framework that included not only the dominant community types of five-needle pines, but also all other forest community types, which have previously been demonstrated to encompass abundant regeneration of five-needle pine species. Our analysis revealed similar trends for whitebark and limber pines: both species exhibited increased levels of mortality that are occurring faster than growth of surviving trees, as well as abundant regeneration in forest types that are not dominated by five-needle pines. Although limber pine has experienced lower mortality rates than whitebark pine, it nonetheless showed signs of decline that are comparable to broad-scale indicators exhibited by whitebark pine 10 years prior. Great Basin bristlecone and foxtail pine mortality rates were relatively low, and their populations exhibited flat diameter distributions except for low recruitment from seedling to sapling size-classes. Our findings suggest that five-needle white pine species would benefit not only from increased seedling recruitment, but also from enhanced recruitment among older and larger age and size classes, both in stands dominated by conspecifics and in stands dominated by species other than white pines.

Whitebark Pine Restoration in BLM Montana's Core Areas

by Emily Guiberson | Rich Byron | Bureau of Land Management | Bureau of Land Management

Keywords: , cone collection, daylighting, monitoring, planting, prescribed fire

Bureau of Land Management Montana/Dakotas State Forester, Rich Byron, and Dillon Field Office Forester, Emily Guiberson, highlight the restoration actions being implemented in Montana's whitebark pine core areas. They will discuss ongoing projects, agreements, and plans to establish a statewide monitoring protocol, with the intent of making it standardized within the agency. In addition, future conservation strategies and objectives will be outlined throughout the presentation.

Assisted Migration of Whitebark Pine to Higher Latitudes and Elevations near its Northern Limits

by Sybille Haeussler | Linda Tackaberry | Hugues Massicotte | UNBC | UNBC (retired) | UNBC (retired)

Keywords: assisted migration, mycorrhizae, northern BC, planting trials, whitebark pine

Here, we report 5- to 10-year results of three assisted migration field trials established from 2011 to 2013 near whitebark pine's northern limits in British Columbia. These trials (1) tested whether absence of compatible mycorrhizal fungi could inhibit whitebark pine migration beyond its current range, and (2) compared performance of provenances from northwest Washington to northern BC and Alberta, outplanted at 848 - 1934 m elevation across a range of minimally to severely disturbed sites.

Although we did not examine mycorrhizae after seedlings left the nursery, we found no evidence that mycorrhizal colonization level inhibited field performance. In our 2011 field trial, adding 250 ml of soil collected beneath mature whitebark trees to planting holes had no effect on 10-year field performance of seedlings grown in standard nursery potting mix and outplanted on moderately to severely disturbed sites (all $P > 0.10$). In our 2012 and 2013 trials, mycorrhizal development in the nursery was excellent in soils collected from alpine (above current whitebark elevation range) and subalpine (within current elevation range) habitats. Seedlings grown in alpine soils had larger diameters than seedlings grown in subalpine soils when they left the nursery ($P = 0.001$) and retained this advantage for at least 5 years in the field ($P = 0.05$).

ANOVA results indicated that effects of provenance on field performance of whitebark pine seedlings were either non-significant or subtle (diameter growth of one southern BC provenance was ~2 mm less than that of other provenances). Detailed statistical modeling incorporating climate parameters and seed weight is warranted once 10-yr data are available from all sites.

Environmental conditions at local planting sites (a disturbance severity gradient in the 2011 trial; an elevation gradient in 2012-13 trials) substantially affected seedling performance. (1) Seedling growth rates increased with disturbance severity with growth on severe wildfire and burned clearcut sites exceeding that on unburned clearcuts or beneath mountain pine beetle snags. (2) Blister rust infection rates (60% on the severe wildfire site) also increased with disturbance severity, creating a difficult trade-off between growth performance and infection risk. (3) Seedlings suffered more damage from extreme weather at alpine or montane elevations than at subalpine elevations. (4) Seedlings in thin soils over bedrock or moist productive soils had lower survival and more damage than those in submesic soils.

These early results support current BC guidelines allowing whitebark pine seed transfer over broad geographic and elevation ranges. They suggest that, in an uncertain climate, selecting genotypes with high rust resistance should be prioritized over fine-tuning provenances. Longer evaluation is needed to determine how provenance affects seed crops and the other ecosystem services provided by whitebark pine. Planting above treeline and over a broader spectrum of microsites must proceed cautiously. Research to support enhanced seedling resistance to disease and climate stress (e.g., through fungal inoculation in the nursery, or by planting trees where they can form fungal linkages with mature trees) should be explored further at these northern whitebark pine limits.

High-Elevation Five-Needle Pine Seedling Traits Vary According to Climatic Gradients

by Lacey Hankin | Sarah Bisbing | Elizabeth Leger | Department of Natural Resources, University of Nevada - Reno | Department of Natural Resources, University of Nevada - Reno | Department of Biology, University of Nevada - Reno

Keywords: Great Basin bristlecone pine, adaptive traits, climate change, limber pine, restoration, whitebark pine

Tree species' persistence potential under rapid, ongoing climate change will depend on their capacity to adapt, migrate, or acclimate via phenotypic plasticity. The variation in and distribution of seedling traits conferring early success is a key knowledge gap for restoration and conservation of forest ecosystems. We quantified variation in seedling traits and evaluated evidence for local adaptation and plasticity in five-needle pine populations across the Great Basin using paired greenhouse and field common gardens. We specifically asked: a) how do seedling traits vary within- and among-populations across environmental gradients? and b) to what extent do populations exhibit local adaptation and/or plasticity to growing conditions? We planted seeds from three whitebark, four limber, and four Great Basin bristlecone pine populations in all source locations and in field-collected soils in the greenhouse and assessed germination, growth, and survival. We used linear mixed effects models to evaluate trait variation and quantified plasticity to soils using a plasticity index.

Our findings highlight interspecific seedling strategies for allocating biomass, yet high intraspecific trait variation across broad climatic gradients. Bristlecone showed significantly higher specific leaf area and root length but lower growth rates and total biomass. Whitebark invested relatively more in root mass when sourced from drier conditions, while limber reduced fine root production but limber and bristlecone produced faster and more growth when sourced from warmer and drier conditions, consistent with interspecific niche differentiation. Notably, limber populations germinated at significantly higher rates in both the greenhouse and field gardens regardless of conditions, suggesting potential advantages of this generalist species over co-occurring bristlecone pine under increasingly stressful environmental conditions. Significant population- and species-level differences in trait responses across climatic gradients and plasticity to soils suggest that successful restoration of climate change-impacted forests may hinge upon species-soils interactions and selection of climate-informed seed sources.

Potential Changes in Climate Suitable Habitat of Whitebark Pine in Greater Yellowstone Under Climate Scenarios

by Andrew J Hansen | Montana State University

Keywords: climate change, environmental tolerances, whitebark pine

Whitebark pine (*Pinus albicaulis*) (PIAL) is a proposed threatened species that plays a keystone ecological role in the Greater Yellowstone Ecosystem (GYE). Its population response to climate change is of high interest to managers because climate-induced declines may adversely affect critical ecosystem services that this species provides. While previous studies of reproductive size classes of the species have projected dramatic reductions in area of suitable habitat under climate warming scenarios, it has been suggested that the species can tolerate warmer and drier conditions if seedlings and saplings are not competitively excluded by other conifer species. Thus, we asked if juvenile-sized PIAL are found in warmer and drier locations than larger individuals, under the assumption that competitive exclusion would require several years to decades to influence the distribution of regenerating PIAL. We used a new genetic technique to distinguish non-cone bearing PIAL from the more warm-dry tolerant limber pine (P. flexilis) among samples collected along transects extending from lower treeline to the subalpine around the GYE. We discovered that smaller diameter PIAL were not proportionally more abundant at lower elevations, suggesting that competitive exclusion may not be the primary mechanism limiting this species' low elevation distribution. In contrast, the small size class PIAL was slightly less warm-dry tolerant than larger individuals. This suggests that the zone of regeneration of PIAL has shifted upwards in elevation in recent decades, perhaps associated with the observed warming in the GYE. In comparison to a previous study of reproductive-sized trees (>20 cm dbh) from a coarser (1.6 km) sampling frame, however, the predicted zone of suitable habitat of PIAL (<1 cm dbh) was 122 m lower in elevation. We conclude that consideration of the fine-scale distribution of PIAL near lower treeline suggests that the tree species is slightly less sensitive to climate warming than found by previous studies of reproductive-sized trees, but, nonetheless, large range contractions of PIAL in GYE are likely under projected future climates.

Summer Air Temperature for the Greater Yellowstone Ecoregion (770–2019 CE) Over 1,250 Years

by Karen Heeter | Maegen Rochner | Grant Harley | University of Idaho | University of Louisville | University of Idaho

Keywords: blue intensity, climate change, paleotemperature, tree rings

Annual surface air temperatures across the western United States (US) have increased by more than 1°C since ca. 1900. Continued warming will likely lead to increased drought conditions and exacerbated fire regimes, threatening to push ecosystems past their natural tolerance limits and potentially lead to a major demographic shifts, especially in subalpine ecosystems. Millennial-length paleo-temperature proxies derived from tree ring data in North America are rare, but extremely important because they enable comparisons between the modern warming trend and conditions that existed during the Medieval Climate Anomaly (MCA; ca. 950–1250 CE). We use latewood blue intensity from high-elevation *Picea engelmannii* to reconstruct late-summer maximum air temperature for the Greater Yellowstone Ecoregion (GYE) spanning 770–2019 CE. Using a robust regression model (r^2 = 0.60), the 1,250-year reconstruction documents regional expression of past warm and cool events, such as an anomalously warm periods spanning several decades of the MCA (1050-1070 CE), a prolonged warm period spanning the fifteenth to sixteenth centuries, and the Maunder and Dalton minima of the Little Ice Age. We demonstrate that historical temperature trends during key periods across the region—such as the MCA, the sixteenth century, and the LIA—were more variable as compared to those presented at hemispheric- global scales. Summer temperature variability across the GYE shows multi-centennial agreement with trends in solar irradiance, volcanic activity, snowpack, and other regional- to-hemispheric temperature records. Further, we emphasize the need for more millennial- length paleo-temperature records to better understand historical variability of climatic change in subalpine ecosystems of North America.

Fire-Caused Mortality of Whitebark Pine

by Sharon Hood | Alina Cansler | US Forest Service | University of Washington

Keywords: prescribed fire, tree mortality, wildfire

Whitebark pine forests are dependent on fire, yet the individual trees are susceptible to mortality from fire. This can make it difficult to minimize whitebark pine mortality when implementing prescribed burns and managing wildfires for resource benefit. The Fire and Tree Mortality Database contains approximately 1300 whitebark pine tree records collected from Wyoming, Montana, Idaho, and Oregon. We use the database to examine the likelihood of whitebark pine mortality after fire based on tree size and fire injury level, as well to evaluate the accuracy of whitebark pine mortality models in fire effects and behavior software systems. We also tracked mortality for 20 years after wildfire for a subset of the whitebark pines and report on long-term mortality from both fire and mountain pine beetle. This talk will present data on fire-caused tree injury levels that are likely to cause mortality, model accuracy, as well as discuss ways to mitigate mortality when burning.

Whitebark Pine Conservation Program at Crater Lake National Park: 2003-2021

by Jen Hooke | Crater Lake National Park

Keywords: High-elevation white pine communities in the Pacific West Region

Crater Lake National Park has maintained an active Whitebark Pine Conservation Program since 2003 that supports the Park's goals and objectives for conserving and restoring whitebark pine. Major elements of the program include identification of rust-resistant individuals; annual monitoring and protection of rust-resistant trees; sampling long-term monitoring plots to track trends in health and disease incidence; and conducting restoration plantings and monitoring the efficacy of those plantings. Out of 126 whitebark pines that have undergone screening trials, 41 have shown genetic resistance to blister rust although 7 of those "resistant" trees have since died. The Park's long-term monitoring plots show large whitebark pine continuing to decline from blister rust- and mountain pine beetle-caused mortality. The Park has conducted six restoration outplantings with more plantings planned for the future.

Whitebark Pine Restoration at Crater Lake National Park: The First Ten Years

by Jen Hooke | Richard Sniezko | Crater Lake National Park | USFS Dorena Genetic Resource Center

Keywords: Special session led by Richard Sniezko on resistance/genetics

Six whitebark pine restoration plantings have been established at Crater Lake National Park from 2009-2016 and monitored on an almost-annual basis. These plantings serve a dual purpose as: 1) restoration plantings including seedlings from rust-resistant 'parent' trees to reestablish whitebark pine in areas suffering from high mortality rates; and 2) these plantings also include seedlings from parent trees deemed 'susceptible' to blister rust and therefore double as genetic trials to confirm field resistance to blister rust and to monitor genetic variation to other biotic and abiotic influences. The current plantings serve as focal restoration populations, laying the foundation for further future natural regeneration and spread of resistance. In addition, the current living rust-resistant parent trees identified in Crater Lake National Park will also help spread resistance. The Crater Lake National Park plantings provide one model of the potential use of genetic resistance to begin the restoration of a non-commercial forest tree species. They also provide a conservation education tool to raise public awareness of the potential for restoration using genetic resistance.

Physiological Responses of Whitebark, Limber and Great Basin Bristlecone Pines to Environmental Stress

by Sean Hoy-Skubik | Danielle Ulrich | Montana State University | Montana State University

Keywords: climate, drought, heat, physiology, seedlings, stress

High-elevation five-needle pines are currently facing grave threats as a result of global change. Warming and drying trends are expected to increase the physiological stress these trees face, leading to future increases in the already widely prevalent mortality events. A primary conservation strategy for mitigating these threats is the outplanting of seedlings. However, the success of this strategy is dependent on the selected seedlings' level of resistance, response to, and ability to recover from future climate-type stressors. To investigate these themes, we used a greenhouse-based study to impose heat and drought treatments, followed by a recovery period, on Whitebark Pine (*Pinus albicaulis*), Limber Pine (*Pinus flexilis*) and Great Basin Bristlecone Pine (*Pinus longaeva*) seedlings, using 3-, 4- and 6-year-old individuals from 17 populations from climatically distinct seed sources. We regularly measured leaf gas exchange and predawn and midday leaf water potentials, and we sampled needles, stems, and roots for non-structural carbohydrate concentrations regularly throughout the experiment. These measurements were used to elucidate patterns of stress resistance across species and populations, including different carbon allocation strategies. Historic climate variables from each population's seed source were compared to measures of heat and drought response. The experiment is ongoing and results will be available by September 2021. Our findings serve to improve our understanding of future responses of these trees to climate change induced stress, as well as the viability of conservation strategies such as outplanting.

Great Basin Bristlecone and Limber Pine in Great Basin National Park: First Look into the Long-Term White Pine Monitoring Efforts

by Nicole Hupp | Devin Stucki | Jeff Galvin | Bureau of Land Management | National Park Service, UCBN| National Park Service, MOJN

Keywords: NPS, bristlecone, limber pine, long-term monitoring

Great Basin National Park (GRBA) in north-central Nevada contains two white pine species, Great Basin bristlecone pine (*Pinus longaeva*) and limber pine (*P. flexilis*) forests. Together, these two species occupy over 3,000 ha within the Park. Like all white pines, these treasured forests are at risk of infestation by many threats including white pine blister rust (*Cronartium ribicola*), mountain pine beetle (*Dendroctonus ponderosae*), and dwarf mistletoe (genus Arceuthobium). To understand these risks, ecologists at GRBA, the Upper Columbia Basin Network Inventory & Monitoring Program (UCBN), and Mojave Desert Network Inventory & Monitoring Program created a white pine monitoring protocol. While GRBA specific, this protocol is a rendition of Mckinney et al (2013), which is implemented in three NPS long-term monitoring programs. The purpose of this monitoring project is to document and interpret changes in community dynamics in forests containing *P. longaeva* and *P. flexilis*. To do this, the protocol measures the following variables: Tree species composition and structure; Forest birth, death, and growth rates; Incidence of WBPR and level of associated crown kill; Incidence of mountain pine beetle and severity of tree damage; Incidence of dwarf mistletoe and severity of tree damage; Cone production of white pine species.

This program began in 2018 with a three-year rotating panel design. In 2020, all plots were established and measured once (n=30). Here we present some background on the details of this monitoring program and an initial assessment of the population. The results we share focus on status of white pines in GRBA; we do not yet have enough data to understand trends. We did find encouraging results of the initial sample: white pine blister rust was not present on any species in GRBA. Together, UCBN, GRBA, and MOJN provide an example of a collaborative long-term monitoring program that is designed to provide both short term and long-term information about white pines in GRBA.

Integrating Restoration Practices for Whitebark Pine with Climate Change Distributional Shifts

by Katie Ireland | Andrew Hansen | Robert Keane | Kristin Legg | MTDNRC, Sage Grouse Habitat Conservation Program | Montana State University, Department of Ecology | USDA Forest Service, Missoula Fire Sciences Laboratory | Inventory and Monitoring Division, Greater Yellowstone Network, National Park Service

Keywords: adaptive management, conservation, planning, status

Whitebark pine (*Pinus albicaulis*), a keystone species in subalpine forests of the northern Rocky Mountains, is expected to be vulnerable to climate-mediated shifts in suitable habitat, pests, pathogens, and fire. Adapting whitebark pine restoration practices to changing climates is a critical need for management of these forests. However, few existing climate adaptation frameworks prescribe where to place management actions to be most effective under anticipated future climate conditions. We developed three spatially-explicit management alternatives for whitebark pine in the Greater Yellowstone Ecosystem, including: (1) no active management, (2) current management, and (3) climate-informed management which used projected climate suitability for WBP and competing tree species to place management actions. Using a landscape simulation model, we evaluated the effect of these management alternatives on whitebark pine population under projected future climate conditions on two landscape within the Greater Yellowstone ecosystem. Using the FireBGCv2 simulation model, we evaluated the effect of these management alternatives on whitebark pine populations under projected future climate conditions on two landscapes within the Greater Yellowstone Ecosystem. Climate change impacts on whitebark pine differed amongst the two landscapes. Management actions had little impact on whitebark pine abundance in the Wind River Range of northeastern Wyoming, where increased fire frequency led to loss of competing tree species and large classes of trees, resulting in establishment of whitebark pine seedlings but an inability to mature due to frequent burns. In the Beartooth Mountains of Montana, white pine blister rust was the primary driver of whitebark pine mortality under future climate, so planting blister-rust resistant seedlings has the potential to mitigate some climate-driven loss of whitebark pine forests. These results suggest that climate change impacts on whitebark pine forests will be context- dependent and require restoration strategies customized to local conditions.

Restoring the Crown: A Plan for Whitebark Pine Restoration in the Crown of the Continent Ecosystem

by Melissa B. Jenkins | Anna W. Schoettle | Jessica W. Wright | US Forest Service (retired) | USFS, Rocky Mountain Research Station | USFS, Pacific Southwest Research Station

Keywords: Ecosystem services; range mapping; restoration plan; whitebark pine; Pinus albicaulis; effective population size; Crown of the Continent Ecosystem; reforestation

The Crown of the Continent Ecosystem (CCE) has experienced the highest levels of whitebark pine (*Pinus albicaulis*; WBP) mortality anywhere in the species range with >90% losses in some locations. This dramatic decline is due to a combination of mountain pine beetle (*Dendroctonus ponderosae*), effects of fire suppression, and the nonnative pathogen *Cronartium ribicola*, all of which have the potential to be exacerbated by climate change. The CCE spans 18-million acres in parts of Montana, British Columbia, and Alberta. While internationally recognized for its biodiversity and ecological integrity, the CCE is also very jurisdictionally fragmented which presents challenges when coordinating restoration efforts. This presentation describes a pilot approach done on three jurisdictions within the CCE to: 1) Assess the relative conservation value of WBP habitat with consideration of the individual jurisdiction's mission and mandates, and 2) Prioritize actions needed to ensure the persistence of functional, evolutionarily stable, WBP populations. An example from one national forest is provided where land managers have completed the final step of developing a long-term restoration plan that assigns site-specific restoration treatments. The original pilot approach has been modified based on lessons learned and has recently been used to guide a restoration strategy for both WBP and limber pine (*Pinus flexilis*) across the full CCE; preliminary results of that analysis will be covered briefly.

Whitebark & Worldview-2 - Looking into the Possibilities of Using Satellite Imagery to Distinguish Subalpine Tree Species

by Stephanie Jouvet | Brendan Wilson | Selkirk College | Selkirk College

Keywords: Whitebark, WorldView-2, remote sensing

Whitebark pine has the largest distribution of all the five-needle white pines in North America and can be found throughout western Canada and the United States. The pine is listed as endangered under the Canadian Species at Risk Act due to comprehensive impacts such as white pine blister rust, pine beetle, fire suppression and climate change. Understanding the species' distribution and occurrence will aid in establishing localized recovery strategies to help conserve this keystone species. Remote sensing has streamlined traditional field-based data collection methods, reducing the time and resources needed to map detailed forest information over large spatial extents. The objective of this research was to explore the use of multispectral satellite imagery (WorldView-2) to distinguish between whitebark pine and other two main subalpine species (Engelmann spruce and subalpine fir) within the Darkwoods Conservation Area in British Columbia. An Object-Based Image Analysis and the maximum likelihood classification algorithm was used to classify the WorldView-2 imagery. The mean spectral signatures of the three tree species had similar values across the eight bands available in the imagery, but Engelmann spruce has slightly lower reflectance values in the Red Edge, NIR 1 & 2 bands. The overall producer's accuracy was 68.75% and the Kappa coefficient was 64.29%. Whitebark pine's individual accuracy was disappointingly low at 21%, compared to that of the spruce (68%) and fir (63%). The overall classification accuracies were relatively low compared to other studies conducting tree species classification. This could have been due to several factors including a smaller sample size of geo-referenced trees, the quality of reference data, and classification algorithm used. Future work will include collecting a more extensive dataset and exploring the combination of LiDAR and multispectral imagery, as well as other classification algorithms.

An Assessment of Whitebark Pine Populations in National Parks of the Pacific States

by Erik S. Jules | Phillip J. van Mantgem | Benjamin G. Iberle | Jonathan C.B. Nesmith | Regina M. Rochefort | Humboldt State University | U.S. Geological Survey | Humboldt State University | National Park Service | North Cascades National Park Service Complex

Keywords: Inventory & Monitoring Program, National Parks, Pinus albicaulis, matrix model, mountain pine beetle, white pine blister rust

Whitebark pine (*Pinus albicaulis*) is a long-lived tree found in high-elevation forests of western North America that is currently experiencing population declines across its range. We evaluated monitoring data from seven national parks in the USA, including Sequoia & Kings Canyon, Yosemite, Lassen Volcanic, Crater Lake, Mount Rainier, Olympic, and North Cascades national parks. We first summarized stand structure, presence of blister rust, and mountain pine beetle prevalence for each park. Next, we used a stochastic, size-structured population model constructed using separate, detailed demographic data from Crater Lake to forecast future trends in the parks under conditions of increased beetle activity and ongoing blister rust infection. Blister rust infected 29-54% of whitebark pine in all the parks except the two southernmost parks, Sequoia & Kings Canyon and Yosemite, where infections rates were 0.3% and 0.2%, respectively. Similarly, the proportion of dead trees in the two southernmost parks was low (0-1%), while they ranged from 10-43% in the other parks. Assuming Crater Lake conditions, model projections suggested an average population decline of 27% in the parks over the next century, though we do not expect these declines unless the parks experience marked increases in beetle prevalence and blister rust. Overall, blister rust was a common stressor in most parks we studied, while beetles appear to have impacted whitebark pine mortality most at Crater Lake. If climate change increases future beetle prevalence, then our study also illustrates the potential rates of whitebark pine decline given a scenario of increased pest outbreaks with simultaneous blister rust infections.

Overview of Whitebark Pine Restoration Efforts in U.S. Forest Service Regions 1, 2, and 4

by Ellen Jungck | USFS Northern & Intermountain Regions

Keywords: whitebark, management, planting, cone collection, relase & weed, mountain pine beetle protection

Individual and cooperative efforts with cone collection, planting and survival trends, release & weed and mountain beetle protection treatments, and gene conservation actions will be discussed. The evolution of cone collection and nursery techniques will be discussed, as will the development of sub-regional cooperative groups and associated conservation strategies.

Anatomy Measurements from Whitebark Pine (Pinus Albicaulis) In the Beartooth Mountains, WY Show Strong Seasonality Signal

by April L. Kaiser | Grant Harley | Georg von Arx | Maegen Rochner | University of Idaho | University of Idaho | Swiss Federal Institute WSL | University of Louisville

Keywords: dendroclimatology, paleoclimate, quantitative wood anatomy, whitebark pine

Whitebark pine (*Pinus albicaulis*) populations are declining across southern Canada and northwestern United States. In the Greater Yellowstone Ecosystem (GYE), whitebark pine populations are negatively impacted by increased temperatures, decreased snowpack, and increased disease occurrence. The nutrient-dense seeds from whitebark pine support grizzly bear *(Ursus arctos horribilis)*, Clark's nutcracker *(Nucifraga columbiana)*, and Douglas squirrel *(Tamiasciurus douglasii)* survival and reproductive health. Understanding how whitebark pine respond to variable climate conditions is critical for the continued legacy of the whitebark pine ecosystem and its dependent species. Quantitative wood anatomy (QWA) is a novel dendrochronological method such that anatomical and cellular features are quantified and analyzed in micro-resolution quality. Our study uses QWA methods to measure cellular parameters in whitebark pine from the GYE such as maximum radial cell wall thickness (CWT_{Rad}), hydraulic conductivity (Kh), and lumen area (LA). Whitebark pine CWT_{Rad} has a strong relationship ($p < 0.01$) with maximum temperature both spatially and temporally (1910-2017). Future climatic changes will likely continue to negatively impact whitebark pine populations unless climate mitigation strategies and continued whitebark pine recovery efforts are implemented.

Special Session: The North American Dendroecological Fieldweek (NADEF): Educational and Research Opportunities to Better Understanding 5-Needle Pines

Organizer: Maegen Rochner

Ecology and Distribution of Klamath Foxtail Pine

by Michael Edward Kauffmann | Ecologist

Keywords: Klamath foxtail pine, Pinus balfouriana subsp. balfouriana

California's endemic foxtail pines have established two esoteric populations abscinded by nearly 500 miles of rolling mountains and deep valleys. The species was first described by John Jeffrey in the Klamath Mountains near Mount Shasta in 1852. Later, this species was discovered in the high-elevations (9,000'-12,000') of the southern Sierra Nevada. The ecological context of Klamath foxtail pines (*Pinus balfouriana* subsp. *balfouriana*) in the Klamath Mountains differs drastically from that in the Sierra Nevada due to the divergence of these populations in the mid-Pleistocene (Bailey 1970). Michael Kauffmann has been working to understand the ecology and distribution of this species for nearly 20 years and will provide and update of his work.

Managing Wildfire for Whitebark Pine Ecosystem Restoration in Western North America

by Robert Keane | USFS RMRS Missoula Fire Sciences Lab

Keywords: wilfire, whitebark pine, suppression, fuel treatment, prescribed burning, partial suppression

Wildfire in declining whitebark pine forests can be a tool for ecosystem restoration or an ecologically harmful event. This presentation will detail a set of possible wildfire management practices for facilitating the restoration of whitebark pine across its range in Western North America. These management actions are designed to enhance whitebark pine resilience and health, while also being effective wildfire management measures. The actions are presented by the three phases of the wildfire continuum: Before, during, and after a wildfire. Current pre-wildfire restoration actions, such as mechanical thinning's, prescribed burning, and fuel treatments, can also be designed to be fuel treatment activities that allow more effective suppression of wildfires when needed. Three wildfire strategies can be implemented while the wildfire is burning—full suppression, partial suppression, and wildland fire use (letting some fires burn under acceptable conditions)—for protecting valuable whitebark pine trees and for ecosystem restoration. Finally, post-wildfire activities include planting rust-resistant seedlings and monitoring effects of the wildfires. Recommended wildfire management practices for the wildfire continuum are provided in this presentation.

Effective Actions for Managing Resilient High-elevation Five-Needle White Pine Forests in Western North America under Changing Climates at Multiple Scales

by Bob Keane | USFS Missoula Fire Sciences Lab

Keywords: landscape ecology, Clark's nutcracker, management strategy, whitebark pine, limber pine, restoration treatments, planting

Many ecologically important high-elevation five-needle white pine (HEFNP) forests that historically dominated the high-elevation landscapes of western North America are now being impacted by mountain pine beetle (*Dendroctonus* spp.) outbreaks, the exotic disease white pine blister rust (*Cronartium ribicola*), and altered high-elevation fire regimes. Management intervention using specially designed strategic tactics designed at both range- wide, landscape and stand levels are needed to conserve these keystone species. The goal of this intervention is to promote self-sustaining five-needle white pine ecosystems that have both resilience to disturbances and genetic resistance to white pine blister rust in the face of climate change. Many tools and methods are available for land managers, and in this paper, we summarize possible multi-scaled actions that might be taken as steps toward restoration of these valuable HEFNP forests. Long-term programs, such as inventory, mapping, planning, seed collection, nursery, education, and research provide the materials for effective restoration at finer scales. Stand- and landscape-level passive and active treatments, such as silvicultural cuttings and prescribed fires, in both healthy and declining forests are the foundation of HEFNP restoration and they are described in detail and grouped by objectives, methods, and tactics. And last, there are the proactive tree-level actions of planting and protection that may be used alone or to enhance success of other restoration actions. As with any range-wide restoration effort, the administrative, policy, legislative, and societal barriers to implementation of an effective restoration effort will also be discussed.

The Grizzly-Squirrel-Pine Nut Connection

by Kate Kendall | Ursine Ecological

Keywords: Tamiasciurus hudsonicus, Ursus americanus, Ursus arctos, black bears, caching behavior, feeding ecology, grizzly bears, red squirrels, whitebark pine seeds

In 1977, we knew that whitebark pine seeds were one of many grizzly bear foods in the intermountain west of North America. What we didn't know was the significance of this food source, how grizzlies were able to obtain cones or seeds, and if red squirrels played a role in making them available to bears. In this talk I provide a glimpse into my research and other's that answered these and other questions about the grizzly-squirrel-whitebark pine connection. Highlights include grizzlies digging through 3 m of snow to feed on cones in the spring, black bears with resin-encased club feet from harvesting sticky cones, feeding bears pine nuts in the zoo, and the Boone and Crocket record size squirrel cache of whitebark pine cones.

Whitebark Pine (Pinus Albicaulis) Growth and Defense in Response to Mountain Pine Beetle Outbreaks

by Nickolas Kichas | Sharon Hood | Gregory Pederson | Richard Everett | David McWethy | Montana State University | U.S. Forest Service | U.S. Geological Survey | Salish Kootenai College | Montana State University

Keywords: defense, growth, mortality, mountain pine beetle, resin ducts, whitebark pine

Whitebark pine (*Pinus albicaulis*) is a critical forest species of Northern Rocky Mountain upper subalpine ecosystems, yet little is known about the physiological response of whitebark pine to disturbance (e.g. fire, bark beetles, and pathogens) across a range of diverse environmental gradients. Resin–based defenses have long been recognized as the primary mechanism by which conifers respond to attack by bark beetles and pathogens and several studies have linked resin duct properties to survivorship during periods of increased beetle activity. However, to our knowledge, no studies have compared axial resin ducts in the secondary xylem of whitebark pine across pairs of living and dead whitebark pine trees to better understand survivorship following multiple disturbances including mountain pine beetle and white pine blister rust. We found a clear distinction in growth and defense characteristics between live and dead whitebark pine. Across our study sites on the Flathead Indian Reservation in northwestern Montana, live whitebark pine produced larger resin ducts with a greater annual investment in resin–based defenses than whitebark pine that died. Resin duct size, duct area, and relative duct area were all greater in live whitebark pine (by 56%, 48%, and 57%, respectively) and these were the most important variables influencing whitebark pine survivorship. In contrast, whitebark pine that had died grew faster over time (22% larger ring widths) than their live counterparts and also produced more resin duct structures (20% more ducts on average). Whitebark pine at our study sites exhibit differing strategies in the allocation of resources toward growth and defense, with the majority of survivors of recent disturbance investing more in defensive structures than growth. Our results support the idea that maintaining genetic variability and the associated suite of differing physiological traits promotes diverse response strategies to a complex array of biophysical and biological stressors that might leave a species vulnerable to extinction across its range.

Clark's Nutcracker Habitat Use and Breeding Ecology in a Heavily Impacted Whitebark Pine Ecosystem

by Vladimir Kovalenko | Diana L. Six | Lisa J. Bate | University of Montana | University of Montana | Glacier National Park

Keywords: Clark's nutcracker, Glacier National Park, breeding ecology, foraging ecology, habitat use, migration

This presentation outlines ongoing research in Glacier National Park, focusing on environmental and temporal drivers of Clark's nutcracker occupancy, use of the landscape, migration, and breeding behavior. With high rates of white pine mortality due to white pine blister rust, it is of interest whether the extant whitebark population in the park remains functional at a level that supports a breeding nutcracker population, as a declining bird population is unlikely to sustain natural whitebark regeneration.

The first objective of this study is to inventory whitebark stands throughout Glacier and provide an update on blister rust infection and mortality, as well as live basal area and cone production. Following the findings of previous research, the second objective is to relate these forest characteristics to nutcracker occupancy at the stand level, as well as migration and breeding ecology at the landscape level. We are also interested in whether higher nutcracker occupancy correlates with greater seedling density.

Previous research proposes that there is a threshold level of whitebark cone density below which nutcracker occupancy becomes highly unlikely. Additionally, research in the Greater Yellowstone Ecosystem suggests that in years of low whitebark cone production, birds are more likely to leave the ecosystem in winter and less likely to breed. Due to high white pine mortality, we seek to learn if nutcracker breeding still occurs in Glacier, even in a relatively good cone year. We are also interested in whether nutcrackers are caching sufficient quantities of seeds to remain in the ecosystem year-round; and if not, where they migrate to and whether they return.

Hold Your Fire: Whitebark and Limber Pine Regeneration Is Not Fire-Dependent in the Canadian Rocky Mountains

by Jodie Krakowski | Joyce Gould | Margriet Berkhout | Robin Gutsell | independent consultant | Office of the Chief Scientist, Alberta Environment and Parks, Government of Alberta | Rocky Mountain House Forest Area, Alberta Agriculture and Forestry, Government of Alberta | Species at Risk Stewardship, Alberta Environment and Parks, Government of Alberta

Keywords: fire, forest health, limber pine, regeneration, whitebark pine

Whitebark (WBP) and limber pine (LP) habitats in Canada's Rocky Mountains are moister and cooler than in the USA, with only limited regeneration-fire studies. We examined whether relationships between fire and regeneration in the USA held in Canada. Alberta Wildfire fuel and fire history assessment methods were streamlined to quickly collect key data while monitoring ~250 long-term WBP and LP plots in 2019. Around half (54% of 82 LP and 47% of 106 WBP) stands had no visible fire evidence. Neither latitude nor elevation significantly influenced regeneration density. Regeneration density weakly increased with LP tree density (LP: $R_2 = 0.22$, $p < 0.05$; WBP $R_2 = 0.07$, $p > 0.05$). Fire (presence/absence) did not significantly affect regeneration abundance or health in either species. However, recently burnt (< 20 years) LP stands had significantly less total regeneration (181 sph) compared to unburnt stands (266 sph) or those with older (>20 years) burns (283 sph). In contrast, recently burnt WBP stands had more regeneration (504 sph) than unburnt stands (389 sph) or stands with old burns (165 sph), and also had significantly more rust infection in tall (50 – 140 cm) regeneration than in stands with no or old burns. This may reflect post- fire increases in understory vegetation, including blister rust hosts, especially in wetter sites that support WBP stands, whereas LP is more restricted to dry sites with sparse fuels. Our frequency and presence estimates for older fires are conservative and may have missed historical, unrecorded burns that require dendrochronology to confirm.

Walking the Talk: Alberta's Whitebark and Limber Pine Recovery Program

by Jodie Krakowski | Robin Gutsell | independent consultant | Species at Risk Stewardship, Alberta Environment and Parks, Government of Alberta

Keywords: disease resistance, endangered species, gene conservation, limber pine, recovery, seed orchard,seedlings, seeds, whitebark pine

Alberta established a multi-agency team in 2006 for recovery of whitebark (WBP) and limber pine (LP), listed as Endangered in 2008. The recovery team led data collection and developed recovery plans for WBP (2013) and LP (2014). The team refocused on provincial implementation, submitting a unified plan in 2019 reflecting recent progress. New external team members expand capacity and continue successful cross-jurisdiction information and resource sharing. Core to status and trends assessment is the network of ~250 long-term monitoring plots spanning the Canadian Rocky Mountain region. Spatial inventory and habitat models are published. Gene conservation seed collections include seed from over 200 LP and nearly 50 WBP putatively disease-resistant trees to date, with many in genetic screening. Their seeds produce seedlings for restoration planting of over 15,250 seedlings in over 78 hectares since 2018. Monitoring tracks success to adapt practices. A replicated WBP daylighting project was established in 2018. Rust-resistant WBP and LP seed orchards are in development and species-specific seed zones are established with supporting provenance trials through partnerships. Grafted scion from 30 resistant LP parents are planted in a seed orchard at Waterton Lakes National Park, with more pending to enhance genetic diversity and future seed production. Guidelines and best practices are shared with industry and agency staff for impact avoidance and mitigation, especially for irreplaceable disease-resistant trees and restoration projects. The slow growth, irregular cone crops, and extensive, remote ranges of these species necessitate a long-term commitment among diverse partners to for effective recovery.

Opportunities and Challenges for Uav/Drone Surveys of High-elevation Pine Forest

by Andrew Latimer | UC Davis

Keywords: drone photogrammetry, neural networks, tree mortality

Rates of disturbance, including fire and disease, have been increasing in many high- elevation forests. The effects at stand scale are often detectable from Landsat and other satellite time series of images, yet the effects on individual trees and the dynamics of local spread are too fine to be seen in this imagery. Drone imagery coupled with neural networks for image segmentation and classification is becoming a powerful tool for surveying forest overstories at high resolution and may offer an intermediate-scale solution for surveying disturbance in these forests. We present results from surveys of lower-elevation conifer forests to demonstrate the method and show how accurate these surveys currently are. In mid-elevation mixed conifer forest, we can survey about 40 hectares/day, which corresponds to on the order of 10,000 individual tree crowns. Tree mortality is easy to evaluate from a distance with near-perfect accuracy, while species identification is more challenging. These methods promise to allow rapid surveys of forest composition and structure, and tree disease and mortality in high-elevation forests as well. Challenges, however, include more difficulty identifying individual trees, and morphological similarity between some growth forms the high five pines, as well as access for flights in wilderness areas.

What Have We Learned in 20 Years of Whitebark Pine Restoration in Glacier National Park?

by Rebecca Lawrence | Jennifer Hintz Guse | Glacier National Park | Glacier National Park

Keywords: Glacier National Park, monitoring, restoration

Glacier National Park has been dedicated to restoring five needle pine trees including limber and whitebark pine across the park landscape since 1998. In the ensuing decades we have been studying how to do this successfully and efficiently. Have we been effective in this and what have we learned? Since the start of the 20th century five needle pines in Glacier have been dying from white pine blister rust. Due to their importance as a keystone and foundation species, restoration efforts are critical to their continued survival in the park and for the hundreds of species that rely on the trees. From 2000 to 2020 crews have planted 23,608 whitebark seedlings throughout the park and approximately 21% of those trees have been monitored. 7071 limber seedlings have been planted from 2002 to 2017 on the east side of the park of which 25% were monitored. The data from early monitoring results have influenced subsequent plantings.

The overall survival of monitored whitebark seedlings is 48% and 17% for limber seedlings. This includes plantings in burned and unburned sites, as well as spring, late summer and fall plantings. The most successful plantings are generally in burned areas with success in both spring and fall plantings. Preliminary results suggest inoculation with a mycorrhizal slurry increases survival and growth, particularly when planted in a burn. As summers are getting warmer and drier continued attention to planting locations, exposure, elevation and weather conditions are increasingly important to the survival of planted five needle pine seedlings.

Special Session Organizer: Kristin Legg, kristin_legg@nps.gov

Special Session Title: Five Needle Pine Research, Monitoring, and Restoration in National Parks

Convergence and Conservation of Major Gene Resistance to White Pine Blister Rust in White Pines

by Jun-Jun Liu, Anna W. Schoettle, Richard Sniezko | 1. Canadian Forest Service – Pacific Forestry Centre, Natural Resources Canada, Victoria, BC, Canada. 2. USDA Forest Service, Rocky Mountain Research Station, Fort Collins, CO, USA. 3. USDA Forest Service, Dorena Genetic Resource Center, Cottage Grove, OR, USA.

Keywords: association study, comparative genetic mapping, major gene resistance (MGR), resistanceevolution, southwestern white pine

Major gene resistance (MGR) to white pine blister rust (WPBR) is found in four species of North American native white pines, including *Cr1-Cr4* in sugar pine (*Pinus lambertiana*), western white pine (*P. monticola*), southwestern white pine (*P. strobiformis*), and limber pine (*P. flexilis*), respectively. Their R loci determine phenotypic expression of MGR after infections by corresponding avirulent races of *Cronartium ribicola*. Understanding the molecular basis of phenotypic convergence would assist utilization of resistance resources in conservation and breeding. Previous genetic mapping works compared R loci from sugar pine (*Cr1*), western white pine (*Cr2*), and limber pine (*Cr4*) and mapped them on different *Pinus* consensus linkage groups (LGs). We performed phylogenetic analysis of their positional candidates of nucleotide binding site leucine-rich repeat (NLR) genes, demonstrating that they are not orthologous. Thus, convergence through independently parallel evolution appears to be responsible for the generation of R genes in these three species. In contrast, southwestern white pine has overlapping geographical ranges with limber pine, and hybrid zones between the two species allow gene flow. Using genotyping arrays with *Cr4*-linked SNPs, association mapping of southwestern white pine *Cr3* was performed to examine if MGR trait convergence with limber pine is caused by a shared orthologous locus. Our evidence suggests that conservation of an ancient MGR specificity may be present and shared between the two species, which may reflect an ancient mechanism, presumably without coevolved interaction with *C. ribicola*, through either descent from an ancestral progenitor, or relatively recent introgression from one to another.

Assessment of Cumulative Whitebark Pine Mortality in the Greater Yellowstone Ecosystem

by William W. Macfarlane | Brian Howell | Jesse A. Logan | Chad Garlick | Gabe Henry | Robert E. Spangler | Utah State University | USDA Forest Service | Retired: USDA Forest Service | Utah State University | Utah State University | U. S. Fish and Wildlife Service

Keywords: Cronartium ribicola, Dendroctonus ponderosae, GIS, Greater Yellowstone Ecosystem (GYE), Landscape Assessment System (LAS), Mountain pine beetle, Pinus albicaulis, aerial survey method, climate change impacts, pest monitoring, white pine blister rust, whitebark pine

An aerial survey method called the Landscape Assessment System (LAS) was used to assess mountain pine beetle (Dendroctonus ponderosae; MPB) - caused mortality of whitebark pine (*Pinus albicaulis*) across the Greater Yellowstone Ecosystem (10,232 km2; GYE). This large- scale implementation of the LAS method consisted of 12,300 km of flightlines, along which 4,434 geo-tagged, oblique aerial photos were captured and processed. The Mountain Pine Beetle-caused Mortality Rating System, a landscape-scale classification system designed specifically to measure the cumulative effects of recent and older MPB attacks on whitebark pine, was used to classify mortality. A rating of none to severe (0-4.0 recent attack or 5.0-5.4 old attack) was assigned to each photo based on the amount of red (recent attack) and gray (old attack) trees visible. The approach produced a photo inventory of 74 percent of the GYE whitebark pine distribution by area. For the remaining 26 percent of the distribution, mortality levels were estimated based on an interpolated mortality surface. Catchment level results that combine the photo-inventoried and interpolated mortality indicate that 44 percent of the GYE whitebark pine distribution showed severe old attack mortality (5.3-5.4 rating), 37 percent showed moderate old attack mortality (5.2 to 5.29 rating), 19 percent showed low old attack mortality (5.1-5.19 rating) and less than 1 percent showed trace levels of old attack mortality (5.0-5.09). No catchment was dominated by recent attacks indicating that the outbreak of the early 2000's has ended and that current MPB populations are likely low, but not eliminated. As weather once again becomes favorable for MPB, these endemic populations are once again capable of explosive population growth and if there are available whitebark pine to serve as food future MPB outbreaks in are possible. Spatially explicit mortality information produced from this assessment is efficient for targeting restoration and conservation management at a landscape scale. Given the level of mortality and whitebark pine's important role as a foundation and keystone species, future research aimed at monitoring mortality and understanding the cascading effects of widespread decline of this species on wildlife, hydrology, and forest structure and function is warranted.

Alpine Treeline Ecotones Are Potential Refugia for a Montane Pine Species Threatened by Bark Beetle Outbreaks

by Colin Maher | University of Alaska Anchorage

Keywords: Pinus albicaulis, boundary, climate change refugia, edge, mountain pine beetle, tree mortality, whitebark pine

Warming-induced mountain pine beetle (*Dendroctonus ponderosae*; MPB) outbreaks have caused extensive mortality of whitebark pine (*Pinus albicaulis*; WBP) throughout the species' range. In the highest mountains where WBP occur, they cross alpine treeline ecotones (ATEs) where growth forms transition from trees to shrub-like krummholz – some of which survived recent MPB outbreaks. This observation motivated the hypothesis that ATEs are refugia for WBP because krummholz growth forms escape MPB attack and have the potential to produce viable seed. To test this hypothesis, we surveyed WBP mortality along transects from the ATE edge (locally highest krummholz WBP) downslope into the forest and – to distinguish if survival mechanisms are unique to ATEs – across other forest ecotones (OFEs) from the edge of WBP occurrence into the forest. We replicated this design at 10 randomly selected sites in the US Northern Rocky Mountains. We also surveyed reproduction in a subset of ATE sites. Mortality was nearly absent in upper ATEs (mean ± 1 s.e. % dead across all sites of 0.03 ± 0.03% 0-100 m from the edge and 14.1 ± 1.7% 100-500 m from the edge) but was above 20% along OFEs (21.4 ± 5.2% 0-100 m and 32.4 ± 2.7% 100-500 m from the edge). We observed lower reproduction in upper ATEs (16 ± 9.9 cones·ha_{-1} and 12.9 ± 5.3 viable seeds·$cone_{-1}$ 0-100 m from the edge) compared to forests below (317.1 ± 64.4 cones·ha_{-1} and 32.5 ± 2.5 viable seeds·$cone_{-1}$ 100-500 m from the edge). Uniquely high WBP survival supports the hypothesis that ATEs serve as refugia because krummholz growth forms escape MPB attack. However, low reproduction suggests ATE refugia function over longer time periods. Beyond our WBP system, we propose that plant populations in marginal environments are candidate refugia if distinct phenotypes result in reduced disturbance impacts.

Winter Damage Is More Important than Summer Temperature for Maintaining the Krummholz Growth Form above Alpine Treeline

by Colin Maher | University of Alaska Anchorage

Keywords: Pinus albicaulis, growth limitation, krummholz flags, plant-climate interactions, snow transport, treeline ecotone, whitebark pine, wind

Understanding the processes that control alpine treelines, the elevational limits of tree growth forms, has been a central question in ecology and is growing in importance with concern over climate change. Cool summer air temperatures are currently thought to be the ultimate limiter of upright tree growth at alpine treelines globally. However, winter damage has long been recognized as a shaping force near alpine treelines. Low-growing krummholz growth forms provide an opportunity to test hypotheses about the controls of upright growth in environments above current treelines.

To distinguish between effects of growing season temperature, winter damage and their interaction on preventing upright growth in krummholz, we conducted a field experiment on krummholz growth forms of *Pinus albicaulis* over the summer and winter of 2015-2016 at 10 mountain top sites in the Tobacco Root Mountains, Montana, USA. We experimentally manipulated four factors using a fully crossed design: shoot position (natural low position in the krummholz mat vs. propped up above the krummholz mat), summer warming (warming chamber vs. ambient), winter exposure (shelter cage vs. exposed), and elevation position (local high vs. low krummholz limits). We also conducted an observational study of the climatic conditions associated with recent natural emergent stem establishment from krummholz.

Experimentally propped shoots that were exposed in winter experienced the highest mortality (10-50%), while propped shoots in shelter cages and shoots located within the krummholz mat, whether caged or not, had low mortality (0-10%). Summer warming had little influence on shoot mortality. Surviving mat shoots had marginally higher growth rates than surviving propped shoots during the early growing season after treatments were established. Natural emergent stem establishment was associated with warmer than average summer temperatures, but also warmer winter temperatures, lower winter wind speeds, and lower snowpack.

Synthesis. Our results suggest winter damage plays a more important role than does growing season temperature in maintaining the krummholz growth form. While warming may increase opportunities for emergent shoot establishment above krummholz mats, establishment of upright trees in the krummholz zone will also

Long-Term Assessment of the Efficacy of Prescribed Burning and Mechanical Thinning for Restoration of Whitebark Pine

by Enzo Martelli | Cara R. Nelson | Robert E. Keane | Andrew Larson | Department of Ecosystem and Conservation Sciences, W.A. Franke College of Forestry and Conservation, University of Montana, 32 Campus Drive, Missoula, MT, USA | Department of Ecosystem and Conservation Sciences, W.A. Franke College of Forestry and Conservation, University of Montana, 32 Campus Drive, Missoula, MT, USA | USDA Forest Service, Rocky Mountain Research Station, Missoula Fire Sciences Laboratory, Missoula, MT | Wilderness Institute and Department of Forest Management, University of Montana, 32 Campus Drive, Missoula, MT, USA

Keywords: ecological succession, mechanical thinning, natural regeneration, prescribed burning, restoration, tree mortality

Whitebark pine (*Pinus albicaulis*), an ecologically important tree of high-elevation ecosystems of western North America, is declining across most of its range due to the combined effects of an invasive pathogen, episodic native insect outbreaks, climate-change- induced increases in wildfire frequency and severity as well as successional replacement due to fire suppression. Concern over these threats has led to the listing of whitebark pine as an at-risk species under both the US and Canadian Endangered Species Acts and an increase in restoration and conservation activities. Prescribed burning and thinning have been proposed as primary strategies for restoration. However, there is only limited available information on the efficacy and effects of these treatments, and projects that have been evaluated have shown highly variable results, suggesting that monitoring is a priority need. We used a 20-year, replicated before-after-control-impact (BACI) study to assess the effects of prescribed burns and mechanical cuttings on rates of whitebark pine mortality, natural regeneration, and successional replacement by shade-tolerant conifers in the northern Rocky Mountains of the United States. Rates of whitebark pine mortality (plot percentage) were 9% lower and 4% higher than control plots for thinned and burned plots, respectively. Natural regeneration (ind./acre) was 25% lower and 30% higher than control plots for thinned and burned plots, respectively. Basal area (sq. m./ha) reduction of competing conifers was of 22% and 25% higher than control plots for thinned and burned plots, respectively. Since the implementation of this experiment in the mid-90's, whitebark pine communities have shown a significant increment in mortality of mature trees and reduced natural regeneration in addition to the treatments effect. These results indicate that active management is an effective strategy to mitigate the effect of current threats, where trade-off's includes promoting natural regeneration at the cost of greater tree mortality.

Clark's Nutcracker Conifer Resource Use in Yellowstone National Park: A Hierarchical Distance Sampling Approach

by Thomas McLaren | Diana Tomback | Nels Grevstad | Walter Wehtje | Doug Smith | Lauren Walker | University of Colorado Denver | University of Colorado Denver | Metropolitan State University Denver | Ricketts Conservation Foundation | Yellowstone Center for Resources | Yellowstone Center for Resources

Keywords: Clark's nutcracker, Yellowstone National Park, foraging ecology

Clark's nutcrackers (*Nucifraga columbiana*) harvest and cache seeds from many western conifer species, contributing to tree regeneration. Multiple seed resources within a region are required to sustain nutcracker populations but especially seeds of whitebark pine (*Pinus albicaulis*), a preferred but declining species. Many conifer species are known to display high variability in annual cone production, resulting in variable food resources for seed foraging species. These variations in seed resource availability are thought to be an important factor in nutcracker habitat use, in part, due to their energy sensitive foraging behavior.

In this study, we examined how nutcrackers use conifer forest community types within Yellowstone National Park to determine all potential seed resources. In 2019 and 2020, we established eleven transects in five different forest community types including whitebark pine, lodgepole pine, Engelmann spruce, limber pine and Douglas-fir. Each transect consisted of five point-count stations, and used distance sampling methods, relative cone abundance indices, and behavioral observations to determine habitat and seed resource use. Using annual cone abundance indices collected along our transects, we analyzed how cone production varies between conifer species and across years. To determine which forest community types in the park are used by nutcrackers within a season and how variation in cone production influences their use within and across years, we applied a hierarchical approach to habitat use modeling using distance observations to estimate detectability within each forest community type in a regression modeling framework.

Ecological Integrity of Whitebark Pine on National Forests in California

by Marc Meyer | Michele Slaton | Shana Gross | Ramona Butz | Carol Clark | USDA Forest Service, Region 5 Ecology Program | USDA Forest Service, Region 5 Remote Sensing Laboratory | USDA Forest Service, Region 5 Ecology Program | USDA Forest Service, Region 5 Ecology Program | USDA Forest Service, Region 5 Remote Sensing Laboratory

Keywords: California, ecological integrity, national forests, whitebark pine

Whitebark pine (*Pinus albicaulis*) forest ecosystems in California are diverse and unique, yet their current status and condition are uncertain. Using a combination of geospatial and monitoring plot data, we assessed the ecological integrity of whitebark pine ecosystems primarily on national forests throughout the state of California. We found whitebark pine ecosystems to be structurally, compositionally, and functionally distinct among subregions of California, and all subregions displayed some evidence of declining ecological integrity. Whitebark pine forests in northern California exhibited signs of greater stand densification (Cascade-Klamath), potential encroachment by shade-tolerant conifer species (Cascade- Klamath and Warner Mountains), and increased tree mortality associated with mountain pine beetle outbreaks (Warner Mountains) than elsewhere in California. Whitebark pine stands in the Sierra Nevada showed signs of stand densification (central Sierra) and localized mountain pine beetle outbreaks (southern Sierra east). Notwithstanding these negative signs, much of the state's whitebark pine ecosystems on national forestlands appear to be relatively healthy and intact compared to more northern latitudes. Active management may be required to restore and build adaptive capacity in California's whitebark pine ecosystems with declining integrity.

Special Session title: High-elevation white pine communities in the Pacific West Region Special Session organizer: Jonny Nesmith

Population Structure and Genotype-Environment Associations of Whitebark Pine in the Sierra Nevada

by Elizabeth R Milano | Amy G. Vandergast | Phillip van Mantgem | Joan Dudney | Jonathan Nesmith | Harold Zald | U.S. Geological Survey | U.S. Geological Survey | U.S. Geological Survey | UC Berkeley | USDA Forest Service | USDA Forest Service

Keywords: Sierra Nevada, drought response, genomic population structure, genotype environment association, whitebark pine

Whitebark pine is a long-lived keystone species in subalpine ecosystems of western North America. Though the species has experienced rapid decline due to threats associated with a changing climate, whitebark pine in the Sierra Nevada has shown a less dramatic overall decline. The Sierra Nevada is the southern range limit of the species distribution and whitebark pine in this region show a variable response to specific biotic stressors, making these populations worthy of further investigation. Our genomic study is part of a multidisciplinary effort to investigate the response of whitebark pine in the Sierra Nevada to a major abiotic threat, extreme drought. Specifically, our aim was to measure genomic diversity and characterize the neutral and selective forces that contribute to that variation. We estimated population structure to identify the pattern of neutral variation within and between occurrences. We then used redundancy analysis, a genotype-environment association analysis, to identify specific genetic loci associated with drought indicative climatic predictors. We found whitebark pine occurrences to be well connected in the Sierra Nevada with a moderate trend of isolation-by-distance and slightly elevated diversity in the south. We also found that 1% of the total genetic variation was significantly associated with a select set of climate variables. Of the 1,270 outlier loci identified, 39% were most associated with climate water deficit, 22% with minimum temperature, 20.5% with actual evapotranspiration, and 18% with annual precipitation. These results lay a groundwork for further investigation into genotype-phenotype associations and genotype-by-environment interactions and provide a set of climate associated loci for future monitoring efforts.

Whitebark Pine Recruitment Following Timber Harvest

by Randy Moody | Ken Wright | Moody Tree | Moody Tree

Keywords: Whitebark pine, silviculture, timber harvest

On many sites whitebark pine (*Pinus albicaulis*) recruitment is closely linked with disturbance, which provides suitable conditions for early seral species, such as whitebark pine, to establish and gain a competitive advantage on late seral species which are often more rapidly growing, and more shade and competition tolerant. Fire is the most frequent stand initiating disturbance resulting in whitebark pine stand initiation, and prescribed fire is commonly applied across the landscape to restore whitebark pine ecosystems.

Like fire, timber harvest presents an opportunity to re-initiate whitebark pine on the landscape as it creates early seral conditions suitable for whitebark pine recruitment. The openings created by timber harvest are often viewed as a potential restoration opportunity; however, the post harvest objective is typically to reforest a site with commercial species to support future timber harvest. The objective of this study was to identify the levels of whitebark pine recruitment into harvest areas and determine if recovery gains were present. We surveyed regeneration levels of all species in five harvest areas ranging in age from five to 26-years. Whitebark pine was present in at all sites with densities ranging from 15 to 1435 stems/ha; rust levels were generally low ranging from 0-18% infection. Although whitebark pine densities were high and rust levels low, at no site was whitebark pine common in the tallest cohort and generally only represented up to 26% of the stems on a site. Although harvest areas provide good sites for whitebark pine recruitment, the trajectory to taller size classes appears unlikely. We recommend reduced commercial species planting in areas with advanced whitebark pine regeneration or thinning activities to improve site conditions and to improve the probability of advancing trees to mature size classes.

Whitebark Pine Tree Improvement and Operational Management on the Helena-Lewis and Clark National Forest

by Tanya E Murphy | Matthew O Voigt | USDA Forest Service | USDA Forest Service

Keywords: cone collection, daylight thinning, management, planting, restoration, scion, seed orchard, test plantation, tree improvement, whitebark pine

Whitebark pine (*Pinus albicaulis*) is present in a pure or mixed conifer stand condition on approximately 4 percent of the Helena-Lewis and Clark National Forest. Although relatively small in quantity, management of whitebark pine has been at the forefront of the silviculture program of work for the last 25 years. The Forest actively participates in whitebark pine tree improvement as part of the Northern Region Genetic Resource Program and Intermountain Whitebark Pine Restoration Program with the objective of producing seed with improved white pine blister rust resistance. The Forest manages the Adams Creek Central Montana seed zone whitebark pine seed orchard, two performance test plantations, and 88 plus trees. Operational management, such as whitebark pine planting, daylight thinning and prescribed fire, has also been implemented to assist with restoration of this valuable species. The Northern Region, State and Private Forestry Forest Health Protection, and the National Forest Foundation have funded the Forest's restoration efforts. Find out what it takes to manage a seed orchard and 88 plus trees, establish and monitor performance test plantations, and collect cones, scion, and aeciospores to fulfill Northern Region's tree improvement mission, as well as implement restoration activities across a 2.9-million-acre land base.

How to Log Whitebark Pine Stands

by Michael Murray | BC Forest Service

Keywords: Harvest retention, silviculture, windthrow

Knowledge of post-harvest survivorship and factors that promote successful retention of whitebark pine in mixed-conifer forests is lacking. The objective is to describe the temporal attrition of retained mature whitebark pine trees and to identify factors that likely influence survivorship during the critical initial post-harvest period. Five separate harvest units in southeastern British

Columbia were assessed. Dendrochronological investigation revealed that retained trees experienced high annual mortality rates (3-16%) across harvest sites during the initial five-year post-harvest interval. By eight years post-harvest, retention survivorship ranged 17-80%. After eight years post-harvest, mortality rates became low. The preponderance of fallen stems oriented towards the northeast suggests that storm system events arriving from the Pacific Ocean are the most significant drivers of blowdown. Survivorship is positively associated with shorter tree heights and longer crown lengths, lower frequencies of disease cankers, greater presence of rodent wounding, and higher numbers of surrounding retained trees. There was little effect based on slope and aspect or distance to forest edge. As an endangered species, harvest operations should practice cautiously when in associated forests. I recommend carefully selecting retention trees, ensuring an adequate number of neighbor trees, and positioning retention patches to avoid predominant storm wind directions.

The Whitebark Pine Genome Project

by David Neale | WPEF

Keywords: reference genome

The primary goal of the Whitebark Pine Genome Project is to develop genetic marker technology that can be used to measure and monitor genetic diversity in whitebark pine populations undergoing restoration. Genetic markers of several types have been used in whitebark pine for many years, but for the most part these markers interrogate non-coding, and thus selectively neutral, regions of the genome. Genetic markers are urgently needed to measure white pine blister rust resistance and adaptation to the environment. For this requirement, genetic markers from protein-coding regions of the genome are needed. Such genetic markers can only be developed following deep transcriptome or full genome sequencing. Transcriptome resources do exist for whitebark pine, but a complete genome sequence has not been produced. The specific objectives of the Whitebark Pine Genome Project are:

1. Sequence, assemble and annotate a complete genome sequence.
2. Identify genes determining white pine blister rust resistance and genes determining adaptation to the environment. This will be done through a combination of genome wide association studies (GWAS) and environmental association analysis (EAA).

Once objectives 1 and 2 are completed, there will be a set of genetic markers that can be used in applied programs to measure and monitor white pine blister rust and adaptation to the environment in populations undergoing restoration.

Effective Monitoring and Evaluation of Whitebark Pine Restoration Treatments—Key to Adaptive Management and Future Applications

by Cara R Nelson | Diana Tomback | Bob Keane | University of Montana | University of Colorado | US Forest Service

Keywords: adaptive management, evaluation, monitoring, restoration, success

Whitebark pine restoration remains, to a large extent, experimental, and we are still exploring the methods to achieve desired conditions. This is especially true with respect to the effectiveness of prescribed burning and silvicultural treatments in meeting treatment goals. To date there is only limited information on outcomes from these treatments and results indicate mixed success. Although monitoring project success and applying outcomes to improve future project implementation are essential aspects of the restoration process, monitoring is often not included in projects or is not done with a sampling design that allows inference. Here, we cover basic approaches to evaluating project success including implementation, efficacy, and effects monitoring, as well as design considerations for each category. Specifically, we discuss when the use of reference stands or models is required, and why sampling both treated and control stands before and after treatment is necessary to determine whether the treatment had an effect. We will also discuss how to determine required replication of monitoring plots and methods for evaluate the efficacy of a monitoring design to avoid over- or under-sampling. Finally, we propose to establish a monitoring network to assist managers with developing and implementing monitoring programs for whitebark pine restoration and for sharing lessons learned. The overall objective is to motivate managers and scientists to increase their engagement in monitoring activities and to provide some tools to improve monitoring practice.

Monitoring the Effects of Prescribed Burning on Whitebark Pine

by Cara R. Nelson | University of Montana

Keywords: burning, mortality, prescribed fire, restoration

Whitebark pine (*Pinus albicaulis* Engelm.) is a keystone species of high-elevation ecosystems that provides hydrologic regulation, nutrient-rich food for wildlife, and unrivaled scenery for wilderness enthusiasts. However, this important tree and the goods and services it provides is in decline, due to the combined effects of altered fire regimes, climate change, white pine blister rust (Cronartium ribicola) and mountain pine beetle (Dendroctonus ponderosae). Despite widespread interest in using prescribed burning as a restoration tool, there is little information on the ecological effects of these treatments at the stand or individual tree level. We were awarded funding through the FHP program to assess the effects of prescribed burning on whitebark pine ecosystems by sampling understory plant dynamics, seedling density, tree growth and mortality, and rates of insect and disease infection at two 5-ha sites in the Mission Mountains of Montana using a before-after-control-impact design. Trees in both the control and treated stands had high rates of blister rust infection prior to treatment (ca. 50%). Pre-treatment sampling occurred during summer 2013 and 2014, the burn was implemented in fall 2014, and post-treatment sampling occurred in summer 2015 and 2016. During the two-year period after burning, 40% of mature whitebark pine trees in the treated stand died (31% in year 1 and 7% in year 2). Nearly 20% of the mature whitebark in the treated stand were scorched up into their canopies; all of these trees died. Another 20% of mature trees in the treated stand had bole scorch up to two meters in height; 60% of these trees died. Three trees that did not have any evidence of scorch also died; two of these showed evidence of blister rust infection prior to the burn. In comparison, there was no mortality of whitebark trees in the control stand. We plan to continue to monitor the Mission Mountain sites, and hope to adaptively add additional prescribed fire sites, in order allow for broader spatial and temporal inference about the effects of prescribed fire on whitebark ecosystems.

Whitebark Pine in Yosemite, Sequoia, and Kings Canyon National Parks: Assessment of Stand Structure and Condition

by Jonathan Nesmith | Micah Wright | Erik Jules | Shawn McKinney | USFS PNWRS | Humboldt State University | Humblodt State University | Fire Sciences Laboratory, Rocky Mountain Research Station

Keywords: Inventory & Monitoring, Sierra Nevada, mountain pine beetle, white pine blister rust, whitebark pine

Whitebark pine is a foundational species in many subalpine ecosystems and has experienced severe population declines. Here we present results on the status of whitebark in the southern Sierra Nevada of California, collected as part of the National Park Service Inventory and Monitoring Program. We selected random plot locations in Yosemite, Sequoia, and Kings Canyon national parks using an equal probability spatially-balanced approach. Tree- and plot-level data were collected on forest structure, composition, demography, cone production, crown mortality, and incidence of white pine blister rust and mountain pine beetle. We measured 7,899 whitebark pine, 1,112 foxtail pine, and 6,085 other trees from 2012–2017. Whitebark pine occurred in nearly-pure krummholz stands at or near treeline and as a minor component of mixed species forests. Less than 1% of whitebark pine were infected with white pine blister rust and <1% of whitebark pine displayed symptoms of mountain pine beetle attack. Whitebark pines in the southern Sierra Nevada are relatively healthy compared to other portions of their range where population declines have been significant and well documented. However, increasing white pine blister rust and mountain pine beetle occurrence, coupled with climate change projections, portend future declines for this species, underscoring the need for broad-scale collaborative monitoring.

Special session title: From genes to tree-rings: characterizing long-term climate drivers and extreme drought effects on Sierra Nevada whitebark pine

Special session organizer: Joan Dudney & Elizabeth Milano

Status of Whitebark Pine: Where We Go from Here

by Amy Nicholas | Sean Sweeney | Lisa Solberg Schwab | Ben Solvesky | Julie Reeves | Doug Keinath | USFWS | USFWS | USFWS | USFWS | USFWS | USFWS

Keywords: endangered species act, proposed rule

On December 2, 2020, the U.S. Fish and Wildlife Service (Service) published a proposed rule (85 FR 77408) to list the whitebark pine (*Pinus albicaulis*) as a threatened species under the Endangered Species Act of 1973, as amended (16 U.S.C. 1531 et seq.) (Act). Included in the proposed rule is a special rule pursuant to section 4(d) of the Act that identifies actions necessary to conserve and recover the whitebark pine, as well as a limited number of prohibited acts (85 FR 77408). While the proposed 4(d) rule would not relieve federal agencies of their consultation obligations under section 7 of the Act, it includes exceptions that allow for optimal, flexible, and adaptive forest management activities conducted by Federal agencies that can advance whitebark pine conservation. Designation of critical habitat (CH) was deemed not prudent for the whitebark pine, hence CH was not proposed. Should the proposed listing become finalized in early 2022, the Service will take further steps including but not limited to: 1) Initiation of recovery planning in concert with external partners currently developing a range-wide Restoration Plan and 2) Development of a programmatic consultation(s) for the species in order to streamline the consultation process for project-related actions that may affect whitebark pine.

Whitebark Pine Regeneration Densities, Niche, and Dynamics Three Decades after the 1988 Yellowstone Fires

by Elizabeth Pansing | Diana Tomback | American Forests | University of Colorado Denver

Keywords: microsite, regeneration, wildfire

Although wildfires provide whitebark pine regeneration opportunities and prevent successional replacement by faster growing conifers, increasing fire frequency and severity threaten the persistence of this imperiled species. The 1988 Yellowstone fires, harbinger of future fire regimes under climate warming, provided an opportunity to characterize the post-fire whitebark pine regeneration niche and timeline to canopy closure. In 1990, 275 permanent plots (20m2), stratified by relative moisture availability (xeric, mesic), were established at two study areas, Henderson Mountain (HM) and Mount Washburn (MW), in areas that experienced stand-replacing fire during the 1988 Yellowstone fires and in adjacent unburned forest. We surveyed for whitebark pine regeneration during seven remeasurements between 1990 and 2017 and determined recent stand age-structure, annual regeneration density, growth rates, common microsites, and annual survival rates.

Whitebark pine regeneration density increased with time and differed with moisture availability. Although regeneration densities were similar over time at HM, densities increased more quickly in mesic than xeric sites at MW. Most recent regeneration densities ranged from 0.05 to 0.4 and 0.05 to 1.0 stem/m2 at HM and MW, respectively. At MW, median density was higher in mesic (0.45 stems/m2, IQR: 0.25, 0.65) than xeric (0.10 stems/m2, IQR: 0.05, 0.1) sites, whereas median densities were similar across sites on HM (~0.10 stems/m2). Stand age structure in xeric sites was uniform, whereas mesic sites were dominated by individuals that germinated 5-10 years after fire. Higher seedling growth rates were more likely in mesic than xeric and burned than unburned sites. Regeneration occurred in microsite types including open sites, vegetation, near deadfall, and near other regenerating conifers; common substrates included duff, woody debris, and ash/soot. Differences in regeneration density, growth rates, and microsite types potentially may be explained by topography, microclimate, and increasing water deficits caused by increasing temperature.

Does Drought Response Change with Parental Ancestry across a White Pine Hybrid Zone?

by Lulu Peach | Northern Arizona University / Washington State University

Keywords: adaptation, genetics, hybridization, models, pines

Hybrids of *Pinus strobiformis* (PIST) and *P. flexilis* (PIFL), two high-elevation, five-needle white pine species, have genetic variation associated with climatic gradients across the southwestern U.S. Our goal was to explore the relationship between hybrid index (percentage of ancestry conferred to a hybrid offspring by PIST) and tree-ring-based growth responses to historical drought (or, dendrophenotypes) in order to (1) draw connections between hybrid index, climate-related variables, and tree-ring sensitivity, (2) test for a relationship between four selected dendrophenotypes and hybrid index, and (3) develop multivariate models to predict dendrophenotype based on hybrid index and a variety of environmental covariates. To develop dendrophenotypes, we dated and measured tree rings on cores from 104 trees at 9 geographic sites from Texas to Colorado. We used transformed raw ring measurements to calculate and used these calculations, hybrid indices, and environmental covariates to address our three objectives We found (1) strong, positive correlations between monsoon index (defined as the percentage of annual precipitation received in summer) and hybrid index and between monsoon index and tree-ring sensitivity, as well as significant correlation between hybrid index and other water-availability gradients in the hybrid zone. In addition, we observed (2) weak, positive, but non-significant relationships between two dendrophenotypes (Rc and W:D) and hybrid index and (3) no strong predictive relationships from our final multivariate models. Future research on the effects of hybrid inheritance on growth during drought periods should consider using a larger sample size including more hybrids with greater PIFL ancestry and available single- nucleotide polymorphism (SNP) data in a genotype-phenotype analysis (GPA) in order to improve understanding of future drought responses and management options.

Whitebark Pine Restoration on the Flathead Indian Reservation: Tribal Cultural Benefits Reaped from New Management Strategies

by ShiNaasha Pete | Confederated Salish and Kootenai Tribal Forestry

Keywords: Confederated Salish and Kootenai Tribal Forestry, Whitebark Pine

Today on the Confederated Salish and Kootenai Indian Reservation in Northwest Montana, a conifer species called Whitebark pine is suffering from ramifications of insect and disease, fire suppression, and climate change. Whitebark pine is a keystone species that grows in the upper subalpine and timberline forests of their surrounding mountains. Losing this tree will have a negative effect on animal and plant habitats, contributing an imbalance to different ecosystems below. The Tribe will lose not only a "First Food" of the people (a plant species with a long tradition in the diet and culture), but the Reservation's land and water quality will alter from the loss of different species, mountain snowpack, and erosion. This study explores how the Confederated Salish and Kootenai Tribal Forestry have been restoring the proposed endangered species Whitebark pine across the suitable habitats on their reservation. The Tribes' efforts include scouting appropriate locations for collecting cones, harvesting seeds for planting, and testing for genetic resistance against the fungus called White pine blister rust; which is annihilating Whitebark pine stands in the northwest. The Tribe collaborates with researching groups, and other tribal natural resource entities, applying silviculture practices, and helping create healthy sustainable forest. The findings show that the Whitebark Pine Restoration is successfully moving forward, but more importantly the Confederated Salish and Kootenai Tribes are reconnecting to a part of their ancestral teachings and history by restoring this first food. The results show with continued focus in conserving Whitebark pine across the Reservation and sharing the cultural history between plant, animal, and man; it protects a future for this threatened species, the Salish and Kootenai culture, and the future generations to come.

Does Interannual Availability of Cones Affect Clark's Nutcracker Caching in Burns and Habitat Selection

by Vernon S. Peters | Darren s. Proppe | Janae L. Vriend | Evan K. Buist | Kaleigh S. Greidanus | The King\'s University | St. Edward\'s University | The King\'s University | The King\'s University | The King\'s Univesity

Keywords: Burns, Caching, Limber, Nutcrackers, Whitebark, acoustic analyses

Whitebark pine (*Pinus albicaulis*) and limber pine (*Pinus flexilis*) are endangered species that rely on fire for regeneration to varying degrees at their northern range limit. Despite the crucial role that Clark's nutcrackers (*Nucifraga columbiana)* play in 5-needle pine ecology, no direct assessment of their presence and caching activity has been made in Canada. We tested whether nutcracker activity differed between mast (2018) and non-mast years (2019), on a 2009 fire with burned whitebark and limber pine habitat. We examined three sites for each species, at varying distances into the fire to determine whether nutcracker activity varied between unburned seed sources and burned sites. We used Ravenpro and Kaleidoscope acoustic analysis software to quantify nutcracker calls manually via spectrogram analysis, and to build advanced classifiers to automate detection, respectively. Fall caching activity was 27% higher in mast years based on call bursts/hour, and evenly distributed between both limber and whitebark pine habitat. Interestingly, nutcrackers shifted habitat in the nonmast year, and spent disproportionately more time in burned and unburned whitebark pine habitat (190% more than limber pine habitat). During spring seed retrieval, nutcrackers spend disproportionately more time in whitebark pine habitat, regardless of interannual availability of cones. Manual estimates of nutcracker activity were far more conservative than automated detections with classifiers, which had high levels of false-positive detections due to wind and rain. Our findings suggest that nutcrackers cache more seed in mast years, and cache in both habitats, however they use whitebark pine habitat preferentially during seed retrieval periods, regardless of fall caching activity. More surprisingly, nutcrackers are quite active in both habitats in non mast years, despite the apparent lack of 5-needled pine cones for caching. This suggests that nutcrackers populations are maintained interannually at the northern limits of 5-needled pines, despite the lack of other key conifer food sources. A proper understanding of nutcracker use of burned habitat is important to applying fire management in recovery actions for each of these endangered pines.

Assessing Vulnerabilities in the Mutualism between Whitebark Pine and Clark's Nutcracker in Sierra-Cascade Parks

by Chris Ray | Regina Rochefort | Jason Ransom | Jonathan Nesmith | Sylvia Haultain | Taza Schaming | John Boetsch | Mandy Holmgren | Robert Wilkerson | Rodney Siegel | The Institute for Bird Populations | National Park Service | National Park Service | National Park Service | National Park Service | Northern Rockies Conservation Cooperative | National Park Service | The Institute for Bird Populations | The Institute for Bird Populations | The Institute for Bird Populations

Keywords: Clark's nutcracker, N-mixture model, Vital Signs monitoring, foxtail pine, whitebark pine

Dispersal of whitebark pine, a keystone species at high-elevations in western North America, depends on Clark's nutcracker, a seed-caching bird with an affinity for whitebark pine seeds. To the extent that this dependence is mutual, declines in whitebark seed production could cause declines in nutcracker abundance. Whitebark pine is in decline across much of its range due to interacting stressors, including the non-native pathogen white pine blister rust. We combined avian and tree monitoring data from four national parks to investigate whether spatial or temporal trends in Clark's nutcracker corresponded with trends in whitebark pine. We allowed for trophic dependence of nutcrackers on whitebark in temporal models by linking models of nutcracker density and whitebark trend in a Bayesian framework. Spatial models suggested that nutcracker density responded to resources other than whitebark in at least one park. Temporal models showed strong evidence for dramatic decline or fluctuation in nutcracker density concurrent with significant increases in whitebark crown mortality and trees infected with white pine blister rust. Our results suggest that the mutualism between whitebark pine and Clark's nutcracker is vulnerable to disruption in areas where an introduced pathogen threatens these species.

Field-Testing Whitebark Pine Resistance to White Pine Blister Rust: A Cost-Effective Approach to Progeny Testing

by Iain R. Reid | Charlie Cartwright | Richard A. Sniezko | Richard C. Hamelin | Sally N. Aitken | Parks Canada, Jasper National Park, Jasper, Alberta, T0E 1E0, Canada | British Columbia Ministry of Forests, Lands, Natural Resource Operations, and Rural Development, Cowichan Lake Research Station, Tree Improvement Branch, Mesachie Lake, British Columbia, Canada | Dorena Genetic Resource Centre, USDA Forest Service, 34963 Shoreview Road, Cottage Grove, OR, 97424, USA | Department of Forest and Conservation Sciences, University of British Columbia, Vancouver, British Columbia, V6T 1Z4, Canada | Department of Forest and Conservation Sciences, University of British Columbia, Vancouver, British Columbia, V6T 1Z4, Canada

Keywords: common garden, inoculation, resistance, white pine blister rust, whitebark pine

The endangered species whitebark pine (*Pinus albicaulis* Engelm.) is declining mainly due to the introduced pathogen *Cronartium ribicola* J.C. Fisch, causing the disease white pine blister rust. Low levels of genetic resistance to blister rust are present in whitebark pine. Traditional methods of inoculating seedlings to determine rust-resistance have been effective, but are costly and time intensive. Due to the need for resistant material for planting, this presents a bottleneck in the process of restoring heavily infected stands. Here we test an alternative approach to controlled inoculations that could screen large numbers of families. We screened 214 open-pollinated families from 44 provenances from British Columbia (BC), Washington (WA), Oregon (OR), Idaho (ID), Montana (MT), and Nevada (NV) to determine: (1) the effectiveness of natural rust inoculation from *Ribes nigrum* L. in a common garden; (2) family and provenance level resistance to blister rust; and (3) climate variables related to height and rust resistance. Eighty-one of these families were previously screened in Dorena, Oregon using standard artificial inoculation methods. The natural inoculation was effective, with 73% of seedlings displaying stem symptoms of the disease, and 95% showing rust infection. A clear relationship was found between distance from the nearest *Ribes nigrum* plant and severity of blister rust. Linear mixed models were fitted to height and rust data using *ASReml-R* to estimate breeding values, heritability, and among- population differentiation (QST) for rust-resistance. Population differentiation for rust- resistance was moderate, with the highest resistance in provenances from the Cascade Mountains of Washington, and the most susceptible families from the BC Chilcotin and Okanagan regions. Heritability of rust resistance was moderate (0.23). This method of screening could be used at a broader scale to determine families resistant to white pine blister rust without specialized testing facilities and increase the availability of resistant seedlings for restoration.

Special Session Organizer: Richard Sniezko/ Genetics and Rust Resistance

Whitebark Pine Production at the USFS Coeur d'Alene Forest Nursery

by Emily Rhoades | Molly Retzlaff | Aram Eramian | USDA Forest Service | USDA Forest Service | USDA Forest Service

Keywords: greenhouse, production, seedling, whitebark

Whitebark pine (*Pinus albicaulus* Engelm.) is a high-elevation foundation species that acts as a critical food source for many species. Populations are rapidly declining due to White pine blister rust (*Cronartium ribicola* Fischer), Mountain pine beetle (*Dendroctonus ponderosae* Hopkins), wildfire, and successional replacement by more shade-tolerant species. Restoration efforts are key to ensuring that this species does not vanish from the landscape. The USFS Coeur d'Alene Forest Nursery has grown whitebark pine, for restoration purposes, for almost 30 years. In 2021, the nursery will ship 179,000 whitebark seedlings and continues to see an increase in whitebark orders. To meet this growing demand, the CDA nursery is continually modifying the propagation protocol for whitebark to streamline the seed preparation process and maximize production. Each year, small improvements lead to a more efficient use of labor and growing space, and a more uniform crop.

Whitebark Pine as Natural Archive: The Dendrochronological Value of Whitebark Pine and a Call to Action

by Maegen Rochner | Matthew Bekker | Sally Horn | University of Louisville | Brigham Young University | University of Tennessee Knoxville

Keywords: dendrochronology, tree ring, whitebark pine

Long-lived, montane tree species like whitebark pine and Engelmann spruce may eventually cease to exist due to the combination of climate change and exacerbated native and invasive threats. While this loss would entail dramatic circumstances for mountain ecosystems, it would also result in the irreversible loss of valuable climatological and ecological data. The objective of this research is to develop extended whitebark pine and Engelmann spruce tree-ring chronologies with the potential for use in regional analyses of climate and disturbance but also to serve as an example of the potential of these tree species and the need for increased tree-ring based work. From a high-elevation site in the Beartooth Mountains of Wyoming, we collected hundreds of samples from both living and remnant whitebark pine and Engelmann spruce and, using dendrochronological methods, developed two millennial-length, tree-ring chronologies for the species. The final whitebark pine chronology covered the period 765–2016 CE, and the final Engelmann spruce chronology covered the period 754–2017 CE. While further analyses are needed to fully evaluate the climatological and ecological value of these chronologies, results indicate a strong common signal in both chronologies, along with evidence of possible synchronous disturbances, including fire and insects. In addition to these results, we also provide observations gathered over four years of work at the same study site, which we believe may assist efforts to identify and collect remnant samples with the ability to considerably extend tree-ring chronologies. We hope that our work may serve as a guide for future efforts to collect and analyze data from whitebark pine and other species that are, and may be, threatened with extinction. We call for increased contribution of tree-ring data, from whitebark pine especially, but also from other subalpine species, to accessible archives, such as the International Tree-Ring Data Bank (ITRDB), which will make tree-ring data available for current and future scientists.

High Altitude Abies Religiosa Reciprocal-Transplant Test to Estimate Climatic Change Impacts in Monarch Butterfly Reserve

by Cuauhtémoc Sáenz-Romero | Ana Laura Cruzado-Vargas | Roberto Lindig- Cisneros | Arnulfo Blanco-García | Mariela Gómez-Romero | Leonel Lopez-Toledo | Erick de la Barrera-Montpellier | Universidad Michoacana de San Nicolás de Hidalgo | Universidad Michoacana de San Nicolás de Hidalgo | Universidad Nacional Autónoma de México | Universidad Michoacana de San Nicolás de Hidalgo | Universidad Michoacana de San Nicolás de Hidalgo | Universidad Michoacana de San Nicolás de Hidalgo | Universidad Nacional Autónoma de México

Keywords: Abies religiosa, high altitude conifers, climatic change, provenances, frost damage, drought stress, Monarch Butterfly, conservation

Sacred fir *Abies religiosa* distribute at high-elevations (2800 to 3550 m of altitude) at the Mexican Transvolcanic Belt. Some of their best stands host the Monarch Butterfly overwintering populations at the Monarch Butterfly Biosphere Reserve (MBBR), at the border between Michoacán and México state, central-west México. We established a field/common garden provenance reciprocal-transplant tests at three sites at contrasting altitudes (3400 m, 3000 m, and 2600 m). Planted seedlings originated from seed collected in natural stands along an altitudinal gradient (3000 to 3550 m; eleven provenances, one every 50 m of altitudinal difference), inside the MBBR. The objective is to estimate the phenotypic plasticity (growth and survival in response to be growing on a climate different than their origin) of *A. religiosa* seedlings that might enable them to cope either with warmer climates (by translocating to warmer planting sites, at lower altitudes than their seed origin; site at 2600 m) or with colder climates at assisted migration sites (by translocating to planting sites colder to their seed origin, at 3400 m). Preliminary results indicate heavier mortality at the low altitudinal site, during the dry season, apparently due to exceptionally warm winter temperatures and low winter precipitations. That is consistent with our field observations of a deficit of natural regeneration seedlings recruitment. Results support the proposal of shifting altitudinally upwards Sacred fir populations by assisted migration, to maintain healthy trees, as long as possible, by re-coupling them with the climate for which they are adapted, when such climate will occur at higher altitude, due to climatic change.

High-elevation Chilgoza Pine Forests in the Western Himalayas: Survival Threats & Conservation Exigencies

by Rinki Sarkar | Independent Researcher, Himalayan Studies

Keywords: Himalayan ecology, Chilgoza pine, Edible pine nut, Natural regeneration, Nutcracker, Forest conservation

This research endeavor traces the outcomes of an interdisciplinary field study conceived for tracking the socio ecological trajectory of a high-elevation Himalayan pine. The researcher encountered this species accidentally during her exploratory hiking trails in the Kinnaur Himalayas of India. She gathered that the global distribution of these forests, known as Chilgoza (*Pinus gerardiana*), is extremely sparse posing grave biodiversity concerns. This three-needle pine, which has a soft white-pine lineage, is restricted to stressful montane environments occurring sporadically in patches across the Western Hindukush Himalayas. Native inhabitants in the surrounding vicinity access these forests for pine seeds which are consumed as pine nuts. The fury of cone collection for pine nuts, which caught the researcher's eye, was the main motivation behind this study as she pondered over the repercussions for natural regeneration.

Topographical constraints secluded Kinnaur for a long spell. During this period, pine nuts were collected mainly for self-consumption with meagre threats to Chilgoza forests. Speedy development of a highway through the region in the post-sixties, for strategic defence purposes, dismantled barriers to market access triggering the sale of pine nuts on an unprecedented scale. The rare resource tapped from the wild started fetching a high value in national and international markets causing a vicious cycle of destructive and near total harvesting of cones, declining yield and spiralling prices. 90% of the forest plots evaluated post-extraction had no cones on the branches.

With seeds all sold off as pine nuts, threats to natural regeneration seemed intuitively tenable. The researcher's forest surveys corroborated these untoward consequences. Excessive lopping of branches for cones was all-pervasive and the poor status of natural regeneration was evident from the survey data analysis which reflected an unstable community structure characterized by low seedling and sapling counts, grossly insufficient to sustain the existing forest stock. Native perceptions of the resource, captured during field work, invariably projected a sense of myopia and nonchalance about the ecological implications of ruthless resource extraction. The endemic nature of the resource was not even common knowledge. The urgency of conservation action seemed inevitable for reversing these unsustainable trends.

Unfortunately, a concerted conservation drive is yet to kick off and artificial propagation strategies have not produced the expected results due to poor field initiatives. However, the researcher's inadvertent discovery of the large spotted nutcracker (*Nucifrga multipunctata*) in these forest tracts and its presumed ability to disperse seeds of this Strobus pine, nurtured the possibility of managing forests to enhance nutcracker visitations as an immediate, cost-effective strategy for restoring the depleting forest stock. Towards this end, incentives can be designed for leaving cones behind for the nutcracker besides curtailing cone collection altogether on a rotational basis from a few areas to facilitate recovery and regeneration. Managing grazing is also crucial for preventing browsing or trampling of juvenile recruits. The efficacy of these management strategies can be strengthened if forest restoration efforts are pursued with more scientific rigor preferably through community engagement fostered by awareness programs that pillar sustainable extraction techniques as well.

Satellite-Tracking Clark's Nutcrackers to Investigate Space Use In Declining Whitebark Pine Habitats

by Taza Schaming | Teresa Lorenz | Peter Singleton | Jason Ransom | Diana Tomback | Thomas McLaren | Lauren Walker | Northern Rockies Conservation Cooperative | USFS | USFS | North Cascades National Park | University of Colorado, Denver | University of Colorado, Denver | Yellowstone National Park

Keywords: Clark's nutcrackers, whitebark pine, mutualism, space use, habitat use, habitat selection, movement, dispersal, satellite-tracking, management

Managers need information on Clark's nutcracker habitat use and selection, foraging and seed dispersal movement patterns, and long-distance movements, to inform whitebark pine management plans, and to identify and prioritize areas for whitebark pine conservation or restoration, including best locations to plant blister rust resistant seedlings. Our primary objectives are to evaluate (1) What are nutcrackers' seasonal range sizes, habitat use and selection, and emigration patterns relative to habitat type and health? (2) How does nutcracker space use vary between years and regions? (3) How do nutcrackers connect habitats and ecosystems? Evaluating space use by satellite-tracking nutcrackers in both the Greater Yellowstone Ecosystem (GYE) and Washington's Cascades allows for better inference range-wide. Additionally, comparing data over a large habitat gradient in

Washington – in the eastern, central, and western Cascades; high, moderate, and low whitebark pine density, respectively – best allows for accurate regional management recommendations. We began the first ever study to satellite-track Clark's nutcrackers in 2014, and by spring 2020, will have tagged 20 nutcrackers, including four birds in a new collaborative study in Yellowstone National Park. The seven GYE birds transmitted an average of 713 days, and the Cascades nutcrackers continue to transmit. Preliminary results reveal that, both GYE and Cascades nutcrackers, when separately analyzed, moved over an annual home range of approximately 18,500 ha, a significantly larger area than previously documented via radio-tracking. To date, the Cascades' nutcrackers have remained in the same general area, but in one year, the majority of GYE nutcrackers flew to non-whitebark pine habitat in Utah to overwinter. We will continue to analyze our data in 2020, and will develop resource selection function maps, to inform whitebark pine restoration efforts, including best locations to plant seedlings, and to define which habitat qualities are important for nutcracker population stability and resilience under environmental change.

Taking the Long View: Ecology, Condition, and Outlook for the High-Elevation Five-Needle Pines

by Anna W. Schoettle | Kelly S. Burns | Jodie Krakowski | Shawn T. McKinney | Kristen M. Waring | Diana F. Tomback |Marianne Davenport | USDA Forest Service, Rocky Mountain Research Station, Ft. Collins, CO | USDA Forest Service, Rocky Mountain Region Forest Health Protection, Lakewood, CO | Whitebark Pine Ecosystem Foundation of Canada, Edmonton, AB | USDA Forest Service, Rocky Mountain Research Station, Missoula, MT | Northern Arizona University, School of Forestry, Flagstaff, AZ | University of Colorado Denver, Department of Integrative Biology, Denver, CO | University of Colorado Denver, Department of Integrative Biology, Denver, CO

Keywords: whitebark pine (Pinus albicaulis), limber pine (Pinus flexilis), Rocky Mountain bristlecone pine (Pinus aristata), Great Basin bristlecone pine (Pinus longaeva), foxtail pine (Pinus balfouriana), and Southwestern white pine (Pinus strobiformis), compounding stresses, white pine blister rust, regeneration, climate change, forest health, adaptive capacity

Mountain regions are facing seasonal warming rates that are greater than the global land average putting many of the subalpine forest communities of western North America at risk. The high-elevation five-needle pine species, including whitebark pine (*Pinus albicaulis* Engelm.), limber pine (*P. flexilis* James), Rocky Mountain bristlecone pine (*P. aristata* Engelm.), Great Basin bristlecone pine (*P. longaeva* D.K. Bailey), foxtail pine (*P. balfouriana* Grev. & Balf.), and Southwestern white pine (*P. strobiformis* Engelm.) are often keystone or foundation species in these forests and frequently defined the forest-alpine tree line. In addition to direct climate change impacts (e.g. heat, drought), the high-elevation five-needle pine species are threatened by the compounding stresses of the non-native pathogen that causes white pine blister rust, climate-driven bark beetle impacts, and other pests and pathogens as well as altered largescale disturbance dynamics. The novel suite of stressors and their interactions threaten the sustainability and adaptive capacity of the high-elevation five-needle pines and the resilience of the subalpine ecosystems. Each of the high-elevation five-needle pine species differ in their ecological context, population dynamics, life-history traits, and exposure and sensitivity to the stressors that challenge them. We applied an evolutionary ecology perspective to review the ecology, genetics, and forest health condition of the high-elevation five-needle pine species to identify key vulnerabilities of each species to infer possible future outcomes.

Adaptive Monitoring in Action: Reconsidering Design-based Estimators Reveals Underestimation of Whitebark Pine Disease Prevalence in the Greater Yellowstone Ecosystem

by Erin Shanahan/Kathi Irvine | GRYN-NPS/USGS

Keywords: Bayesian hierarchical model, Cronartium ribicola, Pinus albicaulis, imperfect detection, multiple observer, occupancy models, sampling, trend estimation

Identifying and understanding status and trends in dynamic ecological processes requires continual monitoring over a prolonged time period. A reality of field-based, long-term monitoring efforts is the inevitable turnover of field crew members which may affect consistency and accuracy in data collection over time. Thus far, there has been little effort to quantitatively investigate imperfect detection of tree diseases as compared to wildlife and plant community surveys. Here we evaluate observation error when detecting white pine blister rust (*Cronartium ribicola*) infection in whitebark pine tree (*Pinus albicaulis*) populations in the Greater Yellowstone Ecosystem. We rigorously assess the prescribed analytical approach in the original long-term Greater Yellowstone Interagency Whitebark Pine Monitoring protocol. We consider the design-based ratio estimator initially proposed for estimating white pine blister rust prevalence versus a model-based approach that accounts for the sampling design and imperfect detection and allows for infection probabilities to vary over time and space. Specifically, we found that ignoring observation errors led to lower estimated prevalence of white pine blister rust in the general population. We determined that infection probability has increased since 2004 when monitoring was initiated. However, overall prevalence likely has not changed because of the shift towards smaller diameter trees because of the mountain pine beetle outbreak that spanned over a decade starting in the early 2000s. In addition, we highlight the use of Bayesian hierarchical modeling which offers an alternative method for analyzing data for more practical applications in conservation biology. Our assessment underscores the need for continued evaluation of a long-term monitoring program's analytical procedures as statistical methods improve and new tools become available for estimating and explaining patterns in status and trend of ecological indicators.

Application of High-Resolution (1.2 M) Worldview-3 Satellite Imagery to Resolve Treeline Species Composition: Case Study Using Limber Pine in Rocky Mountain National Park

by Laurel Sindewald | University of Colorado Denver

Keywords: Alpine Treeline Ecotone, Limber Pine, Remote Sensing, Satellite Imagery, Treeline, WorldView-3

Limber pine *(Pinus flexilis)* is an ecologically important conifer in Rocky Mountain National Park (RMNP) and elsewhere in Colorado that is of management concern. Limber pine occurs in scattered subalpine stands in RMNP, but its treeline distribution is unknown. Here, we propose to adapt remote sensing methodology recently developed by coauthor Cross for identifying rainforest tree species to conifers in treeline communities. Our goals are to (1) pioneer the application of high-resolution (1.2 m) satellite imagery from MAXAR's WorldView-3 commercial satellite to distinguish conifer species at treeline, and (2) create a library of spectral profiles for treeline vegetation in RMNP. We will collect spectral reflectance data with a field backpack spectroradiometer and use known GIS locations of tree species to do object-based, rule-based classification of WorldView-3 imagery to estimate park-wide treeline distribution of each conifer species. We conducted a preliminary analysis with ENVI's Example-Based Classification workflow to demonstrate that the three treeline conifer species can be distinguished—limber pine, Engelmann spruce *(Picea engelmannii)*, and subalpine fir *(Abies lasiocarpa)*. The K Nearest Neighbors algorithm was used for the classification with 2/3 of the training samples for each species. The remaining 1/3 of samples was used to do a confusion/error matrix in ENVI. A 70.16% overall accuracy with this simplified method demonstrates that the species can indeed be discriminated. We will continue to refine data inputs and classification approach and to determine if a rule- based classification will improve differentiation in treeline conifers. We will also develop a library of spectral profiles for dominant vegetation species, including willows and birch, to validate our classification results and to facilitate future classifications. The proposed remote sensing approach will pilot cost-effective methodology for community surveys at a fine scale and broad extent, particularly useful for regions that are difficult or hazardous to access.

Mapping and Monitoring California's High-elevation Forests: A State-Wide Field and Remote Sensing Campaign

by Michele Slaton | Alexander Koltunov | Shana Gross | US Forest Service Pacific Southwest Region Remote Sensing Lab | UC Davis Center for Spatial Technologies and Remote Sensing (CSTARS); US Forest Service Pacific Southwest Region Remote Sensing Lab | US Forest Service Pacific Southwest Region Ecology Program

Keywords: California, distribution mapping, disturbance detection, health monitoring, remote sensing

High-elevation white pine forests dominate upper treelines of California's Sierra Nevada, Cascade, Klamath, and Great Basin mountain ranges. These forests include whitebark pine, limber pine, western white pine, foxtail pine, and Great Basin bristlecone pine (*Pinus albicaulis, P. flexilis, P. monticola, P. balfouriana, and P. longaeva*, respectively). The US Forest Service Pacific Southwest Region recently published a state-wide distribution map for whitebark pine, based upon field and aerial survey data, and this map serves as a much- needed input for management and research applications. We report on the development of this map and on the accompanying long-term monitoring strategy stratified across different climate zones and topographic settings, and for which the US Forest Service completed the first round of baseline monitoring in 2019.

Secondly, we report on the implementation of a remote-sensing based disturbance detection system to monitor high-elevation forests in California. Because the remote and rough terrain and short growing season of these forests make traditional field or airborne monitoring difficult to implement with the frequency required to track rapid changes at broad scales, such operational remote sensing methods are in great demand. The Ecosystem Disturbance and Recovery Tracker (eDaRT) is a highly automated, broadly applicable disturbance mapping system that processes all available Landsat imagery, detecting change at 8-16 day timestep, and is operated by the US Forest Service Pacific Southwest Region to generate disturbance map products for science and land management applications. We report results for years 2010-2021 of a newly developed method to estimate canopy cover loss using time series of spectral change associated with eDaRT disturbances. We provide an overview of plans for continued implementation of this tool and its potential to improve the accuracy and efficiency of delivery of high-elevation forest change products for researchers and managers.

Genetic Resistance to White Pine Blister Rust, Restoration Options, and Potential Use of Biotechnology

by Richard A. Sniezko | USDA Forest Service, Dorena Genetic Resource Center

Keywords: genetic resistance, durability, stability, usability, restoration, white pine blister rust, biotechnology, Cronartium ribicola

All white pine species in North America are highly susceptible to the white pine blister rust (WPBR), caused by the non-native fungal pathogen *Cronartium ribicola*. WPBR is present within the geographic range of eight of the nine species in the U.S. (and the four species present in Canada), but has not yet been documented in Mexico. WPBR has led to extirpation of white pines in some areas and drastically reduced the incidence of some species in portions of their range. A low frequency of genetic resistance has been documented in eight of our white pine species, with extensive work on foxtail pine (*P. balfouriana*) now underway. The development of populations of trees with genetic resistance, while retaining genetic diversity and adaptability is seen as a fundamental step in restoring white pine species. Will resistance work? Since trees are long-lived organisms and any use for restoration will serve as the progenitors for future generations, it will be key to establish the durability, stability and usability of resistance in each species. Fortunately, we are able to draw on the many decades of experience with western white pine resistance program which provides a guide to answering such questions. Can biotechnology help? Conventional resistance breeding and tree improvement programs are the key to success in finding and deploying resistance to WPBR. However, biotechnology has the potential to aid resistance programs, but only if used in the context of a holistic resistance program. Biotechnology tools may aid in the initial search for candidate trees or may provide avenues to speed up the development of resistance or incorporate unique resistance not currently found in North American white pines. The strategic use of resistance will vary by species and some options will be discussed.

Range-Wide Testing of White Pine Blister Rust Resistance in Foxtail Pine (Pinus Balfouriana)

by Richard A. Sniezko | Angelia Kegley | John Gleason | Robert Danchok | Brianna McTeague | USDA Forest Service, Dorena Genetic Resource Center | USDA Forest Service, Dorena Genetic Resource Center, Oregon, U.S.A. | USDA Forest Service, El Dorado National Forest, Placerville Nursery, California, U.S.A | USDA Forest Service, Dorena Genetic Resource Center, Oregon, U.S.A. | Weyerhaeuser, Centralia, Washington, USA

Keywords: Pinus balfouriana, foxtail pine, genetic resistance, white pine blister rust

Foxtail pine (*Pinus balfouriana*) is a high-elevation white pine species endemic to California. Due to foxtail pine's small and disjunct range and concerns about the potential impacts of a changing climate and air pollution, the IUCN Red List has listed this species as "Near Threatened." Like all white pine species native to the U.S., foxtail pine is very susceptible to the non-native fungal pathogen *Cronartium ribicola* (causal agent of the white pine blister rust disease, WPBR). A trial inoculated in 2015 suggested that foxtail pine may be the most susceptible of all native U.S white pines. New trials sown in 2018 and 2019 at two USDA Forest Service facilities, Dorena Genetic Resource Center (Oregon) and Placerville Nursery (California) will sample much of the range of this species. In addition, subsets of the seedling families will be tested (1) at differing inoculation densities to characterize resistances that may be expressed at different (low to moderate) densities and (2) included in field trials in CA and OR to validate the seedling inoculation results. These trials will provide the first baseline data on WPBR resistance in this species, critical information for foxtail pine conservation efforts. Early results from the 2018 trial indicate foxtail pine is very susceptible, but there is some variation with inoculum density, with some families showing higher or faster mortality at the higher inoculation levels. Germination data for this range-wide collection will provide an update on the potential for ex situ conservation using seed storage.

White Pine Blister Rust Resistance Programs for Alberta Limber and Whitebark Pine Recovery

by Richard Sniezko | Angelia Kegley | Jodie Krakowski | Genoa Alger | Jun-Jun Liu | Center Geneticist/ Tree Breeder, USDA Forest Service, Dorena Genetic Resource Center, Cottage Grove, OR | Geneticist/5-needle Pines & Seed Orchards, USDA Forest Service, Dorena Genetic Resource Center, Cottage Grove, OR | Consultant, Edmonton, AB | Resource Management Officer, Waterton Lakes National Park, Parks Canada, Waterton AB | Molecular Forest Pathologist, Canadian Forest Service – Pacific Forestry Centre, Natural Resources Canada, Victoria, BC

Keywords: disease resistance, limber pine, restoration, seed orchard, white pine blister rust, whitebark pine

Significant advances continue in Alberta to develop rust-resistant populations that support restoration of impacted limber and whitebark pine habitat. Putatively resistant trees from heavily infected stands have been identified. Parks Canada and Alberta are screening 201 limber pine and over 120 whitebark pine parents at facilities in the USA and Canada. Seed collections from more plus trees will add to the parents being tested in future years. Final results are pending for most parents. Multiple Alberta limber pine populations have dominant single-gene (MGR) resistance; tests are underway to confirm the existence of a second putative major gene in one parent from Waterton Lakes National Park. Interim results indicate low to moderate levels of quantitative resistance in Alberta populations of both species; final results are pending. Preliminary results are used to guide establishment of grafted seed orchards with partners. Provisional whitebark pine seed zones are established, and seed orchards are under development among Canadian partners. One limber pine orchard established in 2020 has 111 grafts from 30 selected parents, with planned expansion up to 1000 from 50-100 genotypes to provide diverse restoration material adapted to the southwest Alberta-USA Northern Rockies seed zone. Another orchard and/or clone bank is planned. As seed zones overlap both provinces, Alberta material is incorporated in whitebark pine seed orchard planning in BC, with planting grafts beginning in 2021 at 2 to 3 sites. MGR limber pine seedlings tested at Dorena were planted at 2 southwest Alberta sites in 2018; long-term observation will determine the durability of this resistance.

A Fungal Focus: Ectomycorrhizal Fungi of Whitebark Pine in Interior British Columbia

by Hanno Southam | Natalie Stafl | Shannon H. A. Guichon | Suzanne W. Simard | Department of Forest and Conservation Sciences, Faculty of Forestry, The University of British Columbia, Vancouver, BC, Canada | Ecologist Team Lead, Parks Canada Agency, Mount Revelstoke and Glacier National Parks, Revelstoke, BC, Canada | Department of Forest and Conservation Sciences, Faculty of Forestry, The University of British Columbia, Vancouver, BC, Canada | Department of Forest and Conservation Sciences, Faculty of Forestry, The University of British Columbia, Vancouver, BC, Canada

Keywords: Suilloid fungi, ectomycorrhizal fungi, plant-fungal interactions, whitebark pine

Whitebark pine is entangled in a mutualistic partnership with an assemblage of ectomycorrhizal fungi (ECMF). This relationship is like sunlight (exaggeration?) – whitebark would not grow and survive without these fungi. We need to start digging, to start paying more attention to these fungi – they are major players in whitebark pine ecology, some with a particular affinity to whitebark pine may be endangered by the loss of their host and they may well impact the success of management and restoration. Here we report on a project in the Columbia Mountains of British Columbia that used next-generation sequencing to identify ECMF on the roots of whitebark pine and in the adjacent soils. At the level of genus, the community of fungi in this region shares the same culprits that have been identified elsewhere (e.g. *Suillus, Rhizopogon, Piloderma, Cenococcum, Meliniomyces* (=*Hyaloscypha*) and *Cortinarius*) but below that level it is completely distinct and appears endemic and adapted to a mixed forest type. ECMF on planted seedlings in our study looked like strangers compared to naturally occurring seedlings and this is expected to influence future survival. The composition of ECMF on mature trees was related to tree health and this may have impacts on a tree's resistance to insects and pathogens. This fungal focus is important not just because of the direct links to whitebark pine ecology but also because it expands our frame of reference to understand trees and their functions as systems.

Education and Research through the North American Dendroecological Fieldweek

by James H. Speer | Margot Kaye | Bryan Black | Grant Harley | R. Stockton Maxwell | Christopher Gentry| Indiana State University | Penn State | University of Arizona | University of Idaho | Radford University | Austin Peay State University

Keywords: Greater Yellowstone Ecosystem, dendrochronology, whitebark pine

Forests are changing around the world due to stress from global change. We examined multiple forest sites in the Shoshone National Forest, during North American Dendroecological Fieldweek events in 2014, 2017, 2018, and 2019. We sampled over a thousand trees from 14 sites to reconstruct temperature, streamflow, spruce budworm outbreaks, fire, mortality dates, and regeneration patterns. We used innovative techniques such as blue intensity, stable isotopes, and quantitative wood anatomy, combined with traditional tree-ring research. We were able to reconstruct over 1,000 years of snowpack, stream flow, and temperature. Mountain pine beetle caused massive mortality in whitebark pine between 2008 and 2012 while spruce budworm outbreaks occurred five times from 1751-2017 CE. Outbreaks have been increasing in duration and the current outbreak is the most severe on record. Whitebark pine is establishing further upslope than was previously recorded, resulting in the conversion of high-elevation meadows to young forests. Whitebark pine from above and subalpine fir from below are regenerating underneath lodgepole pine forests, suggesting a contraction of the lodgepole pine zone in the future. We documented a switch from temperature minimum as the dominant climatic parameter in high-elevation tree growth to drought stress around the 1950s and insect outbreaks may be speeding the conversion to new vegetation types. Overall, the long-term perspective from tree rings shows that recent climate, disturbance, and forest regeneration are outside the natural range of variability in the Greater Yellowstone Ecosystem.

Challenges to Limber Pine at Craters of the Moon National Monument and Preserve

by Devin Stucki | National Park Service, Upper Columbia Basin Network

Keywords: Craters of the Moon, dwarf mistletoe, limber pine, national park service

Limber pine (*Pinus flexilis*) is a foundational and pioneer species and an integral part of the sparse woodlands found at Craters of the Moon National Monument and Preserve (CRMO), located in southern Idaho. Limber pine at CRMO can be found growing near the lower elevational limit of the species' range and tree densities across the rugged lava flows are characteristically low with few exceptions. Limber pine trees at CRMO are facing ongoing threats from native and non-native pests and pathogens, as well as drought. Here we present findings from 9 years of monitoring limber pine woodland dynamics within 2500_2 m plots (n=93) at CRMO including stand composition and structure, cone production and seedling regeneration, and incidence of white pine blister rust (*Cronartium ribicola*), mountain pine beetle (*Dendroctonus ponderosae*), and dwarf mistletoe (*Arceuthobium cyanocarpum*). Rates of mountain pine beetle and white pine blister rust infection remain low, but rates of dwarf mistletoe infection have increased over the course of monitoring with over 40% of limber pine trees currently infected. Dwarf mistletoe is the most commonly recorded cause of mortality within monitoring plots, though rates of mortality remain low for limber pine. The rate of limber pine regeneration has increased while the proportion of living limber pine trees has decreased. Rocky mountain juniper (*Juniperus scopulorum*) stem density is low but increasing at a rate about three times that of limber pine. Long-term monitoring at CRMO is part of a broader collaborative white pine monitoring effort shared by four National Park Service Inventory and Monitoring Networks within the Pacific West Region.

Using Audio Recording Devices and Multiseason Occupancy Modelling to Evaluate Clark's Nutcracker Occurrence in Whitebark Pine Habitats

by Lauren Taracka | Dr. Taza Schaming | Dr. Alison Scoville | Dr. Teresa Lorenz | Central Washington University | Northern Rockies Conservation Cooperative | Central Washington University | U.S. Forest Service

Keywords: Clark's nutcracker, Clark's nutcracker ecology and seed dispersal, acoustic monitoring, whitebark pine

Clark's nutcrackers (*Nucifraga columbiana*) and whitebark pine (*Pinus albicaulis*), both native to the subalpine habitat of Washington's Cascade Mountains, are part of a fascinating mutualism: whitebark pine, a keystone species which is declining, provides high energy seeds, which nutcrackers eat and cache. The goal of this pilot project was to survey nutcrackers in whitebark pine habitat with acoustic recording units, to assess how multiple habitat variables impacted nutcracker occurrence, colonization (likelihood of moving into an unoccupied site), extinction (likelihood of moving out of an occupied site), and detectability. We deployed monitors at twelve sites in the Cascades from June through October 2020 to gather audio data (n = 781 days of data collected), then determined daily presence or absence of nutcrackers with Raven Pro software. In addition, we conducted habitat surveys, cone counts, and in-person occupancy surveys. In our model set, we included weather as a predictor of detectability; and date, local-level whitebark pine density and cone presence/absence, and landscape-level ponderosa pine (*Pinus ponderosa*) and Douglas-fir (*Pseudotsuga menziesii*) density, as predictors of colonization and extinction. Additional analyses include determining whether whitebark pine germination rate is associated with nutcracker occupancy, and comparing in-person surveys to monitor detection. Final results are pending. Our project has strong links to restoration goals and management planning efforts. Understanding how nutcracker occupancy relates to whitebark pine cone, stand and landscape metrics is important for managers because whitebark pine persistence on the landscape depends on presence of birds to disperse seeds. Information on nutcracker habitat use, and foraging and seed dispersal movements will help inform whitebark pine management plans by aiding in identifying and prioritizing areas for whitebark pine restoration.

Climate-Smart Conservation: Putting Adaptation Principles into Practice for Whitebark Pine Conservation and Other Management Challenges

by David Thoma | GRYN-NPS

Keywords: adaptive capacity, climate change, climate impacts, exposure, integration, sensitivity, vulnerabilities

Climate-Smart Conservation was developed by the National Wildlife Federation in partnership with multiple government agencies including the NPS, USFWS, and USFS. The Climate-Smart Conservation (CSC) is a guidance tool for natural resource managers challenged with integrating climate considerations into climate adaptation practices. The CSC framework emphasizes four overarching themes: Act with intentionality through linking actions to impacts; Manage for change, not just persistence; Reconsider goals, not just strategies, and; Integrate adaption into existing work. Using whitebark pine conservation as an example, David Thoma and Erin Shanahan will shepherd participants through the key steps and nuances in the CSC cycle. Components of the cycle include identification of the issue, assessment vulnerabilities further divided into exposure, sensitivity, potential impacts, and adaptive capacity, determining management actions and window of management opportunity, and implementing management actions according to policy and law. The CSC structure is applicable throughout the discipline of natural resource management and simplifies a wicked problem into digestible parts.

The National Whitebark Pine Restoration Plan: Restoration Model for the High Five Pines

by Diana F. Tomback | Eric Sprague | University of Colorado Denver | American Forests

Keywords: High Five pines, National Whitebark Pine Restoration Plan, whitebark pine

Federal land management agencies and non-profit organizations have been aware of the decline of whitebark pine (*Pinus albicaulis*) for more than 30 years, culminating in its recent proposed listing as threatened under the Endangered Species Act. In 2016, the Whitebark Pine Ecosystem Foundation and American Forests jointly proposed the development of the National Whitebark Pine Restoration Plan (NWPRP) to the leadership of the U.S. Forest Service, who would oversee and help support the effort. The plan includes as collaborators other principal land management agencies across whitebark pine's U.S. distribution, including the National Park Service, Bureau of Land Management, several Native American tribal governments, and the U.S. Fish and Wildlife Service. The NWPRP is a geographic plan entailing the identification of essential core or high priority areas in need of restoration, representing 20 to 30% of whitebark pine's range within agency units or tribal lands. The underlying premise is that seed dispersal by Clark's nutcracker will, over time, disseminate blister-rust-resistant genotypes beyond the core areas. The selection of priority areas for restoration is based on the biological and ecological criteria used by the U.S. Fish and Wildlife Service to guide their species status assessments and for assessing species recovery. The steps in the NWPRP development included a National Whitebark Pine Summit in 2017, formation of a Liaison Committee in 2018 to provide feedback and guidance to the organizers, and a series of data calls between 2018 and the present to acquire new distributional information, nominated core areas, and proposed restoration projects for these core areas. The nominated core areas and proposed restoration projects will serve to focus agency resources and support NGO fund-raising to restore a critical number of whitebark pine populations to insure the future survival of the species. Several related high-elevation five-needle white pines are also in decline and may require federal listing in the future, unless there is timely strategic management intervention. Here, we propose that the procedure followed in the National Whitebark Pine Restoration Plan could be applied to develop effective restoration plans for other "High Five" pines.

Do Whitebark Pine Regeneration Trends Support Predicted Distributional Shifts under Climate Change? Examples from the GYE

by Diana F. Tomback | Elizabeth R. Pansing | Tony Chang | University of Colorado Denver | American Forests | Conservation Science Partners

Keywords: climate change predictions, post-fire regeneration, whitebark pine

Whitebark pine (*Pinus albicaulis*) is an ecologically important species of upper subalpine and treeline communities in the Greater Yellowstone Area (GYA) and of conservation concern in the region. Studies of climate change impacts to the GYA concur that whitebark pine is highly climate sensitive and will decline at its lower boundary and eventually throughout most of the GYA. We have monitored subalpine forest recovery after the 1988 fires at the lower boundary of the whitebark pine zone in two study areas (2680 m and 2560m) in both mesic and xeric burned study sites per area, tracking mapped whitebark pine individuals through seven remeasurements of 200 plots. Given that the regeneration niche is expected to be more restrictive than the climate niche for mature trees, we asked whether early signs of climate change impact are apparent in patterns of whitebark pine regeneration. PRISM data since 1989 for the Henderson Mtn., Custer Gallatin National Forest, and Mt. Washburn, Yellowstone National Park, study areas show that average annual air temperature increased significantly during this time in both study areas (by 0.09– 0.10 °C yr_{-1}), but average annual precipitation has not changed, indicating a growing moisture deficit. As of our 2016/2017 plot remeasurements, whitebark pine regeneration density across three burned study sites in two study areas has reached 0.005 sites/m_2, but the whitebark pine regeneration density on the fourth (northwest-facing Mt. Washburn mesic, burned) study site is four times greater at 0.023 sites/m_2. Climatic niche modeling based on AR5 RCP 4.5 and 8.5 GCMs indicate that the Henderson Mtn. study sites have a higher probability (~0.115-0.162) of retaining whitebark pine through 2099 than Mt. Washburn (0.006-0.016), but observed regeneration patterns hint at a different future outcome than the niche models predict. The high regeneration density on the Mt. Washburn mesic, burned study site suggests that local conditions may lead to higher germination rates, greater seedling survival, or facilitation by fast-growing conifers, and possibly alter predictions on a local scale.

Seminal Research: Foraging Strategies of Clark's Nutcracker. 1978. The Living Bird 16, 123-161

by Diana F. Tomback | University of Colorado Denver

Keywords: Clark's nutcracker, seminal paper, whitebark pine

As recently as the 1980s, foresters in North America were confused as to how whitebark pine (*Pinus albicaulis*) seeds were disseminated. For example, in *Forest Trees of the Pacific Slope*, 1908, George Sudworth wrote that whitebark pines shed their seeds in September, and "the cones dry out and open slowly in the high, cold situations where this pine grows." My doctoral dissertation, *The behavioral ecology of Clark's Nutcracker (Nucifraga columbiana) in the eastern Sierra Nevada*, 207 pages, was completed in 1977 at the University of California at Santa Barbara. Until this time, little was known about the role of Clark's nutcracker as a seed disperser, and virtually nothing was published about its relationship with whitebark pine. I presented the first chapter "Ecological strategies for year-round utilization of pine seed by the Clark's nutcracker in the eastern Sierra Nevada" at a major ornithological conference and received an invitation to submit the work to *The Living Bird*, the distinguished journal of the Cornell Laboratory of Ornithology. With a simplified title "Foraging strategies of Clark's Nutcracker," the 1978 paper detailed the annual cycle of family group interactions, seed harvesting, and seed caching. Based on five research field seasons, it clearly documented nutcracker caching and seed recovery principally of whitebark pine but also of Jeffrey pine (*P. jeffreyi*), thus showing how additional seed sources within a region are important to support nutcracker populations. The second chapter of my dissertation, which has been more frequently cited, was published in 1982 as "Dispersal of whitebark pine seeds by Clark's Nutcracker: a mutualism hypothesis" in the Journal of Animal Ecology. This paper demonstrated that nutcracker caches led to whitebark pine regeneration, nutcrackers buried more seeds than they required, and no other animal that cached seeds was likely to be as effective a seed disperser. This latter theme was explored in the 1982 *Oecologia* paper by Hutchins and Lanner. The nutcracker-whitebark pine interaction literally became a textbook example of a coevolved mutualism, and citations to our papers appeared in several ecology textbooks in the 1980s and 1990s.

A Tale of Two Pines: Comparing Seedling Morphological and Stomatal Traits between Whitebark Pine and Limber Pine

by Danielle Ulrich | Montana State University

Keywords: ecophysiology, limber pine, seedlings, whitebark pine

Investigating the ecophysiological mechanisms that determine seedling establishment, such as traits that determine how seedlings acquire light, water, and CO_2, can improve our understanding of current forest boundaries and our ability to predict future ones. Whitebark pine (*Pinus albicaulis*) and limber pine (*Pinus flexilis*) are two species with similar growth habit and dispersal mechanism yet contrasting geographic distributions by elevation. Based on current distributions in the Greater Yellowstone Ecosystem, limber pine has been more associated with warm, dry conditions at lower elevations (870-3810m), while whitebark pine has been more associated with cool, wet conditions at higher elevations (1600-3660 m). Comparing this pair of species enables us to isolate ecophysiological traits that may influence seedling establishment and underlie their contrasting distributions. Here, we quantified crown morphological and stomatal traits in greenhouse-grown 5-year old seedlings of whitebark and limber pine. Specifically, we measured needle length, needle width, stomatal density, stomatal length, stomatal area, fascicle density, the ratio of sunlit leaf area to total leaf area (STAR), and leaf mass per area (LMA).

Preliminary results demonstrated that needle length and needle width did not significantly differ between species ($p>0.05$). Limber pine exhibited significantly greater stomatal density and stomatal length than whitebark pine ($p<0.05$). Limber pine exhibited significantly greater fascicle density, lower STAR, and greater LMA ($p<0.05$). Together, these results suggest that limber pine may have greater stomatal control over the uptake of CO_2 and loss of water, and may be more adapted to high light conditions than whitebark pine.

Growth, Drought Response, and Climate-Associated Genomic Structure in Whitebark Pine in the Sierra Nevada of California

by Phillip van Mantgem | Joan Dudney | Elizabeth Milano | Jonathan Nesmith | Amy G. Vandergast | Harold Zald | U.S. Geological Survey | Department of Environmental Science, Policy, & Management, UC Berkeley | U.S. Geological Survey | National Park Service, Sierra Nevada Network | U.S. Geological Survey | USDA Forest Service, Pacific Northwest Research Station

Keywords: Pinus albicaulis, climate change, dendrochronology, stable isotopes, subalpine forests

Whitebark pine (*Pinus albicaulis* Engelm.) has experienced rapid declines and has recently been proposed as threatened under the Endangered Species Act in the United States. Whitebark pine in the Sierra Nevada of California may be faring better compared to other portions of its range, although biotic and abiotic stressors may still affect populations. We present patterns of stem and needle growth in large, disease-free whitebark pine before and during a recent extreme drought across the Sierra Nevada. We interpret growth patterns considering additional data collected on needle isotope discrimination and population genomic diversity and structure. Whitebark pine had positive to neutral growth trends prior to drought, correlated with increasing minimum temperatures. Growth responses to drought measured in terms of basal area increment and needle length remained positive to neutral during the drought. Stem and needle growth patterns were consistent with interannual patterns in needle ⊠13C, which did not generally suggest moisture stress during the drought years. Individual tree growth response phenotypes appear to be linked to genotypic variation in climate-associated loci, suggesting that growth is not only controlled by site level climate differences, but that some genotypes can take better advantage of local climatic conditions than others. We did not find strong geographical differences in drought growth responses, which was congruent with low levels of observed among site genetic variability. We speculate that snowpack losses during the drought may have lengthened the growing season while retaining sufficient moisture to permit growth for most individuals. While sampled whitebark pine may have responded with neutral to increasing growth during the recent drought in the Sierra Nevada, future conditions may be different, with growth declining in response to increased moisture stress or pest and pathogen activity.

Special Session title: High-elevation white pine communities in the Pacific West Region Special Session organizer: Jonny Nesmith

Fire, Restoration and Climate Change in Southwestern Mixed Conifer Forests: Implications for P. Strobiformis

by Kristen M. Waring | Nicholas Wilhelmi | Northern Arizona University | USDA Forest Service Forest Health Protection

Keywords: southwestern white pine; wildfire; white pine blister rust; gene conservation

Southwestern white pine (SWWP; *Pinus strobiformis*) is an important component of mixed conifer forests of the US Southwest and Sierre Madre Occidental of Mexico. In the US Southwest, the range consists primarily of disjunct populations at higher elevations. Density of SWWP has increased in many locations since the late 1800's due to fire suppression and exclusion. However, the incidence of stand-replacing high severity fire has been increasing amid severe drought conditions across the Southwest. Between 2010-2015, we installed 79 permanent plots to monitor the spread of white pine blister rust (WPBR) in Arizona and New Mexico. Wildfires have since burned 21 of these plots, in 2011, 2014 and 2017. Most plots burned at high severity, and conifer regeneration was limited to moderate and low severity burned areas. In addition, individual trees being assessed for genetic resistance to WPBR, including those which have shown resistance, have succumbed to these fires. Updated results from 2021, 10-year plot remeasurements will be presented. Managers will need to balance silviculture objectives such as fire hazard reduction and restoration to historic conditions with projected losses to WPBR to sustain the species through sufficient regeneration.

Interdisciplinary Research in Southwestern White Pine: Results and Management Implications

by Kristen M. Waring | Samuel Cushman | Andrew Eckert | Lluvia Flores-Renteria | Richard Sniezko | Christopher Still | Christian Wehenkel | Any Whipple | Michael Wing | Justin Bagley | Ethan Bucholz | Marja Haagsma | Jessica Hartsell | Cory Garms | Jeremy Johnson | Erin Landguth | Alejandro Leal Sáenz | Mitra Menon | Ehren Moler | Gerald Page | Andrew Shirk | Jared Swenson | Northern Arizona University | USDA Forest Service Rocky Mountain Research Station | Virginia Commonwealth University | San Diego State University | USDA Forest Service Dorena Genetic Resource Center | Oregon State University | 7. Universidad Juárez del Estado de Durango | Northern Arizona University | Oregon State University | Virginia Commonwealth University | Northern Arizona University | Oregon State University | Northern Arizona University | Oregon State University | Northern Arizona 5. USDA Forest Service Dorena Genetic Resource Center University, | University of Montana | 7. Universidad Juárez del Estado de Durango | Virginia Commonwealth University, 10. University of California-Davis | Northern Arizona University | Oregon State University | University of Washington | Northern Arizona University

Keywords: Pinus strobiformis, adaptive traits, climate change, ecophysiology, genetics, white pine blister rust

Southwestern white pine (SWWP; *Pinus strobiformis*) is an important tree species of the southwestern US and Mexico. Southwestern white pine populations are threatened by climate change and an invasive tree disease, white pine blister rust (caused by the fungal pathogen, *Cronartium ribicola*). Rapid climate change is negatively affecting SWWP by increasing heat and drought stress, and thus challenging its ability to adapt. White pine blister rust causes extensive tree decline and mortality in SWWP. The dual threats of climate change and invasive species make forecasting future tree distributions at continental scales an urgent challenge. The goal of our project is to determine how gene movement among populations, adaptation to disease and drought, heritable changes beyond DNA mutations, and a changing environment interact to govern the success of SWWP. We are utilizing and developing tools to help forecast and manage the future of the species, including those from genomics, ecophysiology, common gardens, tree disease resistance testing, and remote sensing to measure drought tolerance, disease resistance and presence and physiological response. Results will be synthesized and included in cutting-edge landscape genomics models to meet our overarching goal. We will present our framework and current project status, including results and management implications. Completed products include landscape genomic models, a species distribution model incorporating climate change, and analyses of adaptive traits, ecophysiological traits, phenotypic plasticity, genomic variation, and genotype-phenotype associations.

Within and between Species Physiological Traits and Stress Resistances of Three High-Elevation Pines

by Chloe Wasteneys | Danielle Ulrich | Montana State University | Montana State University

Keywords: Great Basin bristlecone pine, Pinus albicaulis, Pinus flexilis, Pinus longaeva, high-elevation pine, limber pine, pine ecophysiology, pine physiological traits, pine stress tolerance, whitebark pine

High-elevation pines inhabit exposed mountain tops, providing habitat for other organisms. However, high-elevation pines are declining rapidly due to a suite of factors exacerbated by climate change including white pine blister rust, bark beetle, and competition with faster growing conifers. The primary restoration strategy is outplanting rust-resistant seedlings. The seedling is the most vulnerable developmental stage and seedling establishment can drive species distributions. Understanding the physiological mechanisms underlying seedling establishment, such as physiological traits and stress resistances, is critical for successful restoration and conservation. Here, in greenhouse-grown 5-year old seedlings, we quantified physiological traits and stress resistances in populations from varying climates of three high-elevation pine species: whitebark pine (WBP, *Pinus albicaulis*), limber pine (LP, *Pinus flexilis*), and Great Basin bristlecone pine (GBBP, *Pinus longaeva*). We also measured physiological traits in WBP seedlings of three ages: 2, 3, and 5 years old. We quantified photosynthetic capacity from photosynthetic-CO2 response curves (maximum photosynthetic capacity (Amax), RuBisCO efficiency (Vcmax)), high light tolerance from photosynthetic-light response curves (maximum photosynthetic rate in saturating light conditions (Asat), light saturation point (Qsat)), drought tolerance from pressure-volume curves (water potential at full turgor and turgor loss point), and thermotolerance from chlorophyll fluorescence and electrolyte leakage curves.

Preliminary data show that the three species do not significantly differ in any of the physiological traits measured at this point. However, marginally significant differences in traits were observed between LP populations from contrasting climates and between WBP seedling ages. The LP population from a lower elevation site (1,645 m) exhibited greater mean Asat (11.35 $\mu mol\ m^{-2}\ s^{-1}$) than that from a higher elevation (2225 m, 6.63 $\mu mol\ m^{-2}\ s^{-1}$) (two-way ANOVA, p-value=0.055), suggesting that the low elevation population exhibited greater high light resistance. Two-year old WBP exhibited greater mean Qsat (1385 $\mu mol\ m^{-2}\ s^{-1}$) than both 3- and 5-year olds (870 $\mu mol\ m^{-2}\ s^{-1}$ and 1130 $\mu mol\ m^{-2}\ s^{-1}$, respectively) but was only marginally significantly different from the 3-year olds (p-value=0.054).

Discovery of WBP, MSU,1974-2001: WBP Environment, Stand Establishment/Development, Physiology, and Follow-Up

by Theodore Weaver | MSU Emeritus

Keywords: and follow-up., physiology, stand establishment/development, wbp environment

The presence of Rocky Mountain WBP forests and their qualities was/were essentially unknown in 1970. Exploration of their nature and processes was an exciting activity of the MSU ecology lab and its USFS associates. I list twenty-five resultant papers, 1974-2001. And suggest important extensions/derivatives.

1. We described open woodlands of multi-stemmed trees based on a sample of ten Montana mountain ranges.
2. We described their environments: Their climates in the northern Rockies and throughout the west - -and compared them with those of their Asian cohorts: eg with respect to temperature, moisture, frosts. Their geologic substrates (lime/hard rock/volcanics). Their soils, including the seasonal dynamics of water and nutrients which influence forest physiology more directly than anything remotely sensible. Wind, vapor pressure deficit, fire and logging were neglected.
3. We observed the establishment and development of WBP forests. We measured seed production and its variability among sites and years and reflected on thier establishment. Once established, stand densities fell through time while cover increased. Total standing crops were followed over 500 years and partitioned, for various applications, among leaf, branch, bole, bark (unique), and roots.
4. Tree performance was related to temperature and light. Stand production was related to structure and resource availability. Decomposition rates are contemplated.
5. Human influences from feeding, though trampling, weed introduction and vegetation recovery from human impacts were recorded. We suggest a solution to the blister-rust problem. Soils recover slowly from compaction, rates over 30 years are still measurable.

Clark's Nutcracker Seed Resource Use and Limber Pine Metapopulation Structure in Rocky Mountain National Park

by Tyler J. Williams | Diana F. Tomback | Community College of Denver | University of Colorado Denver

Keywords: Clark's nutcracker, Rocky Mountain National Park, connectivity, foraging ecology, habitat use, limber pine, metapopulation, seed dispersal

This presentation explores the dynamics of two versatile mutualists, Clark's nutcracker and limber pine, in a northern Colorado landscape mostly unaffected by extreme fire events and white pine blister rust infections during and prior to the study period of 2014 to 2016. We first discuss seed resource use by nutcrackers in Rocky Mountain National Park. What are the drivers that cause nutcrackers to use their preferred seed resource of limber pine as well as the alternative seed resources of ponderosa pine and Douglas-fir? Will these alternative seed resources sustain nutcracker populations in the face of future limber pine mortality from blister rust? Secondly, we explore limber pine's metapopulation structure in Rocky Mountain National Park. We examine the distribution and size of limber pine stands as well as how many stands could be connected by the seed-dispersal flights of nutcrackers. We also examine how historical fire regimes may have impacted local extinction events and how larger fires, such as the 2020 Cameron Peak and East Troublesome fires (the largest in Colorado history) may impact extinction rates.

Nutcrackers foraged on limber pine seeds each year of the study. However, limber pine alone did not provide sufficient energy resources for foraging and caching, and nutcrackers used an additional conifer seed resource each year. Temporal variation in patterns of cone production and maturation among the three conifers as well as temporal changes in habitat use by nutcrackers appear to be the best drivers of conifer use. Limber pine and ponderosa pine stands were utilized primarily as seed resources, whereas Douglas-fir was more frequently used as habitat. With potential future limber pine losses, these other seed resources may be essential to sustain a nutcracker population.

Limber pine component populations ranged from 1 - 400 ha in size with inter-population distances of 1 - 36 km. Nutcracker flight distances indicate potentially high limber pine metapopulation connectivity, even with a smaller nutcracker population. Historically, the limber pine metapopulation likely experienced infrequent extinction events. However, large- scale fires (such as the East Troublesome fire) appear to cause significantly higher extinction rates, potentially complicating metapopulation processes.

Forest Structure Twenty Years after the First Whitebark Pine Prescribed Burn in Banff Nation Park

by Brendan Wilson | Stephanie Jouvet | Rob Walker | Jon Stuart-Smith | Selkirk College | Selkirk College | Parks Canada, retired | Parks Canada

Keywords: long-term, monitoring, prescribed fire, recruitment, succession, whitebark

Following a reconnaissance level survey of whitebark pine stands and blister rust infection in the Canadian Rockies, Parks Canada started a conservation program for the pine species in 1998. With the recognition that fire suppression had reduced the amount of available early serial subalpine forest in the Parks system, the initial action was to assess the use of prescribed fire in creating suitable recruitment habitat for this species. Permanent monitoring plots were established in control and burn treatment units on the southern flank of Observation Peak in Banff National Park. After the initial pre-burn measurement of the treatment units, a high intensity fire was applied to the burn treatment in late August of 1998. The treatments were remeasured the following summer and the results confirmed that the fire had achieved almost 100% mortality of all size classes of the tree and shrub species. Twenty years later, we have remeasured the treatment plots and found substantial recruitment of whitebark pine, compared to the spruce and fir species that previously dominated the burn treatment area. The density of all species combined was 2200 stems/hectare, with whitebark accounting for just under half of this value. Greater that 99% of all these different tree species were seedlings or samplings less than 1.5m tall. No blister rust was evident on whitebark pine in either treatment. Understory shrub layers were dominated by huckleberry and grouseberry in the burn treatment, whereas, the control was dominated by a more diverse mixture subalpine fir, common juniper, Engelmann spruce, and grouseberry. Future burns elsewhere and continued monitoring will provide more insights as to the apparent early success of increasing whitebark regeneration on this northern Rockies landscape.

Poster Abstracts

Teaching Students about Whitebark Pine and Fire Using Hands-on Activities

by Ilana Abrahamson | Missoula Fire Sciences Laboratory, Rocky Mountain Research Station, U.S. Forest Service

Keywords: education, whitebark pine

FireWorks is an educational program about wildland fire science that provides interactive, hands-on activities for students in grades K-12. Activities from the Northern Rocky Mountains and Northern Cascades program feature whitebark pine forests along with two other forest types. Educators can focus exclusively on whitebark pine, or they can address all three forest types so students can learn about the diversity and interdependence of forests. Students use age-appropriate activities to learn about the organisms that live in each forest type, how fires typically burn there, and the historical fire regime. For example, young students learn about whitebark pine's historical pattern of "rollercoaster" (mixed- severity) fires by using feltboard materials to illustrate a story, while older students use the FireWorks Encyclopedia (a series of short essays) to learn about the organisms typical of whitebark pine communities. Each year, thousands of students use FireWorks to learn about fire in whitebark pine communities and many other ecosystems throughout the United States.

Challenges of Restoring a Unique Population of Whitebark Pine

by Robin Garwood | Deb Taylor | USFS | USFS

Keywords: planting, rust resistance, whitebark pine

The Sawtooth National Recreation Area has received support from the Forest Health Protection Whitebark Pine Restoration Program for planting seedlings on Railroad Ridge in the White Clouds Mountain Range. This area contains a unique population of whitebark pine based on having a unique allele. Management of this population is challenging due to its very low white pine blister rust resistance. Results of this planting effort will be explained and how it relates to management of this unique area.

Whitebark Pine Friendly Ski Areas

by Mike Giesey | WPEF

Keywords: conservation, education, guided hike, recreation, ski area, whitebark

This poster will provide an overview of the Whitebark Pine Friendly Ski Area Certification Program. Ski areas represent an excellent opportunity to improve public awareness of whitebark pine and the environmental challenges it faces. The intent of this certification program is to:

- Increase awareness among ski area employees and patrons of issues surrounding whitebark pine
- Guide ski areas in their efforts to conserve and restore whitebark pine
- Provide an opportunity for ski areas and their patrons to be involved in restoring whitebark pine by becoming directly involved in education, conservation and restoration efforts or through monetary donations
- Recognize ski areas that are leaders in whitebark pine education, conservation and management
- And, ultimately, preserve and manage for whitebark pine so that high-elevation recreationists can enjoy the many benefits of whitebark pine

Genetic and Remote Sensing Approaches to Identify White Pine Blister Rust Infection in Southwestern White Pine

by Jeremy Johnson | Marja Haagsma | Richard Sniezko | Gerald Page | Christopher Still | John Selker | Kristen Waring | Prescott College | Oregon State University | USDA Forest Service | Oregon State University | Oregon State University | Oregon State University | Northern Arizona University

Keywords: Hyperspectral Remote Sensing, Pinus strobiformis, genetic resistance, white pine blister rust

Southwestern white pine (SWWP, *Pinus strobiformis*), is a high-elevation, large, long-lived conifer native to the U.S. and Mexico that is susceptible to white pine blister rust (caused by the non-native fungal pathogen, *Cronartium ribicola*). SWWP has a suite of strategies (including Major Gene Resistance and Quantitative Disease Resistance), occurring at low population frequencies, for resisting the fungus. Even though resistance occurs in SWWP, it can be very difficult to identify trees with a resistance mechanism that could be included in breeding for restoration and reforestation. In this study we tested the ability of hyperspectral imaging, combined with a custom motion-control system and machine learning, to identify and track the progression of the disease in order to identify seed sources of potential genetic resistance. We conducted a greenhouse study on 175 open- pollinated seedlings from 10 seed sources selected across the latitudinal range of the species. Seedlings were randomized and half were artificially inoculated with *C. ribicola* spores while the remainder were used as controls. The seedlings were manually scored for disease symptoms and patterns of growth. A support vector machine learning approach was able to automatically detect infection with a classification accuracy of 87% (kappa = 0.75) over 16 image collection dates. Additionally, hyperspectral imaging was shown to accurately detect health vigor status (as a proxy for disease progression) using the normalized photochemical reflectance index, which is a proxy for photosystem II function. This fast and (semi-) automatic approach suggests a way forward to scale up early disease detection in forestry.

Identifying Patterns of Blister Rust Resistance in Southwestern White Pine

by Jeremy Johnson | Richard Sniezko | Kristen Waring | Christian Wehenkel | Prescott College | USDA Forest Service | Northern Arizona University | Universidad Juarez del Estado de Durango

Keywords: Pinus strobiforms, White pine blister rust, genetic resistance, quantitative disease resistance

Southwestern white pine (*Pinus strobiformis*), is a high-elevation white pine native to the U.S. and Mexico and is susceptible to white pine blister rust (caused by the non-native fungal pathogen *Cronartium ribicola*). The species has a major gene, discovered at low frequency in some populations, that conveys complete resistance to the disease but may be overcome in the future by virulent strains of the pathogen. Quantitative disease resistance has also been documented in the species. Even though resistance occurs in southwestern white pine, little is known about the type, frequency, and geographic pattern of resistances across its range. As part of a large collaborative and interdisciplinary study we present early results from a range-wide assessment of 446 families from 104 populations trials sown in 2014, 2016 and 2017, including the first tests of populations from Mexico. Seedlings were artificially inoculated with *C. ribicola* spores and scored for disease symptoms and patterns of growth. All populations show high infection; however, early results indicate that there is a suite of resistance mechanisms that can be leveraged for restoration and reforestation. The incorporation of, physiological, genetic, and remote sensing approaches are improving the speed and precision at which we can identify resistant versus susceptible trees and will aid in the selection, breeding and deployment of resistant candidates for reforestation and restoration. Field trials based on screening results have begun.

Pinus Flexilis Is a Highly Susceptible Host of Dothistroma Septosporum in Canada – First Report

by Jodie Krakowski | Renate Heinzelmann | Tod Ramsfield | Andy Benowicz | Richard Hamelin | Colin Myrholm | consultant | Department of Forest and Conservation Sciences, University of BC | Canadian Forest Service - Northern Forestry Centre, Natural Resources Canada | Alberta Agriculture and Forestry, Government of Alberta | Department of Forest and Conservation Sciences, University of BC | Canadian Forest Service - Northern Forestry Centre, Natural Resources Canada

Keywords: Dothistroma septosporum, limber pine, pathogen

After health surveys suggested *P. flexilis* and other *Pinus* species were infected with *Dothistroma* sp. At the provincial genetic research and conservation facility in Smoky Lake, Alberta, further assays were conducted to confirm the agent. The facility is 350 km northeast of the natural range of *P. flexilis*. Field phenotypic foliar assessments of a germplasm conservation and provenance-family field test, laboratory cultures on solid malt extract-agar (MEA) and in liquid V8 medium, and genomic ITS region sequencing all confirmed *P. flexilis* is a susceptible host of *Dothistroma septosporum* (syn. *Mycosphaerella pini*). This is the first confirmed infection of *D. septosporum* on endangered *P. flexilis* in Canada. Timely reports of potential field observations are important to characterize the extent and to develop appropriate management strategies for this pathogen.

The Great Basin Five-Needle Pine Proactive Strategy Engagement: A Collaborative and Science-based Approach

by Anna W. Schoettle | Maria Newcomb | David A. Charlet | Duncan Leao | Annabelle Monti | Kelly S. Burns | USDA Forest Service, Rocky Mountain Research Station, Ft. Collins, CO | USDA Forest Service, Intermountain Region Forest Health Protection, Ogden, UT | College of Southern Nevada, North Las Vegas, NV | USDA Forest Service, Humboldt-Toiyabe National Forest, Sparks, NV | USDA Forest Service, Humboldt-Toiyabe National Forest, Sparks, NV | USDA Forest Service, Rocky Mountain Region Forest Health Protection, Lakewood, CO

Keywords: Pinus longaeva, Pinus albicaulis, Pinus flexilis, Great Basin, white pine blister rust, Proactive management, management-research partnerships

Great Basin bristlecone pine, limber pine, and whitebark pine protect watersheds, play important ecological functions, are symbols of perseverance and longevity, and are valued by the public. These high-elevation five-needle pines of the Great Basin face direct and indirect effects of climate warming and the threat of the continued spread of *Cronartium ribicola*, the non-native fungus that causes the lethal disease white pine blister rust. The Great Basin ecosystems are unique, and the scope of the threats requires landscape-scale solutions. The primary objective of this ongoing project is to coordinate a cross-boundary partnership of managers, researchers, and professionals to identify critical information gaps and outline priorities for building a strong science foundation that can assist in timely decision making for managing and conserving the high-elevation five-needle pines for resilience to white pine blister rust in a changing climate. We brought together representatives from the USDA Forest Service (R&D, NFS, FHP), USDI Park Service, USDI Bureau of Land Management, USDI Fish and Wildlife Service, academia, and State, Tribe, and non-governmental organizations for a field workshop in July 2021 in eastern Nevada to share knowledge and build a collaborative High-5 group in the Great Basin (figure 1). This project was supported in-part by the USDA FS Region 4 - Rocky Mountain Research Station BeSMART Program intended to spur Intermountain Region (R4) management – FS research partnerships to bring innovative approaches to management challenges.

Preparing for Invasion: Rust Resistance in Limber, Great Basin Bristlecone, and Rocky Mountain Bristlecone Pines

by Anna W. Schoettle | Angelia Kegley | Richard A. Sniezko | Detlev Vogler | Kelly S. Burns | Gretchen Baker | Jeff Connor | USDA Forest Service, Rocky Mountain Research Station, Ft. Collins CO | USDA Forest Service, Dorena Genetic Resource Center, Cottage Grove, OR | USDA Forest Service, Dorena Genetic Resource Center, Cottage Grove, OR | USDA Forest Service, Pacific Southwest Research Station, Placerville, CA (retired) | USDA Forest Service, Rocky Mountain Region Forest Health Protection, Lakewood, CO | USDI Park Service, Great Basin National Park, Baker, NV | USDI Park Service, Rocky Mountain National Park, Estes Park, CO (retired)

Keywords: Pinus aristata, Pinus flexilis, Pinus longaeva, Proactive intervention, genetic resistance, white pine blister rust

Great Basin (GB) bristlecone pine, Rocky Mountain (RM) bristlecone pine, and limber pine are high-elevation five-needle pines threatened by the non-native pathogen *Cronartium ribicola* that causes the disease white pine blister rust (WPBR). The pathogen continues to spread, and the infection front is now in the Southern Rocky Mountains and Great Basin. WPBR is increasing in these areas on limber pine and RM bristlecone pine as well as whitebark pine; GB bristlecone is the only North American five-needle pine that has not been found in the field with the disease. Because these populations are still healthy, they offer the opportunity to assess the baseline frequency of WPBR resistance traits in largely naïve populations, make comparisons among the species, and speculate on the evolutionary origins of resistance traits. Information on genetic resistance traits and their frequencies also supports development of proactive interventions to increase the frequency of resistance in populations before pathogen invasion to mitigate future impacts and sustain ecosystem function during pathogen naturalization. We present results from artificial inoculation common garden studies to assess qualitative and quantitative resistance traits in seedling families for limber and the bristlecone pine species. All three species are highly susceptible to the disease though they differ in their response to inoculation. Variation in disease progression and mortality following inoculation among seedling families and source areas was observed. Disease resistance was evident in some families of each species and varied among species. Limber pine is more easily infected than either of the bristlecone species under common garden conditions. Within the ecological context of each species, genetic resistance to WPBR is likely to be a significant determinant of the species' population trajectory in the presence of *C. ribicola*.

Climate Correlates of White Pine Blister Rust Infection in Whitebark Pine in the Greater Yellowstone Ecosystem

by David Thoma/Erin Shanahan | GRYN-NPS

Keywords: Cronartium ribicola, Greater Yellowstone Ecosystem, Pinus albicaulis, relative humidity, white pine blister rust, whitebark pine

Whitebark pine, a foundation species at tree line in the Western U.S. and Canada, has declined due to native mountain pine beetle epidemics, wildfire, and white pine blister rust.These declines are concerning for the multitude of ecosystem and human benefits provided by this species. An understanding of the climatic correlates associated with spread is needed to successfully manage impacts from forest pathogens. Since 2000 mountain pine beetles have killed 75% of the mature cone-bearing trees in the Greater Yellowstone Ecosystem, and 40.9% of monitored trees have been infected with white pine blister rust. We identified models of white pine blister rust infection which indicated that an August and September interaction between relative humidity and temperature are better predictors of white pine blister rust infection in whitebark pine than location and site characteristics in the Greater Yellowstone Ecosystem. The climate conditions conducive to white pine blister rust occur throughout the ecosystem, but larger trees in relatively warm and humid conditions were more likely to be infected between 2000 and 2018. We mapped the infection probability over the past two decades to identify coarse-scale patterns of climate conditions associated with white pine blister rust infection in whitebark pine.

Maintaining *Pinus strobiformis*, a Tree Species Threatened by Climate Change and White Pine Blister Rust

by Kristen Waring | Richard Sniezko | Nicholas Wilhelmi | Gregory Reynolds | Jeremy Johnson | School of Forestry, Northern Arizona University | USDA Forest Service, Dorena Genetic Resource Center | Forest Health Protection | Forest Health Protection | School of Forestry, Northern Arizona University, USDA Forest Service, Dorena Genetic Resource Center, Department of Environmental Studies

Keywords: Cronartium ribicola, Pinus strobiformis, climate change, genetics, southwestern white pine, white pine blister rust

Climate change and invasive species pose significant threats to forest health and resilience. Southwestern white pine (SWWP; *Pinus strobiformis*), native to the southwestern US and Mexico, faces increasing pressure from hotter, drier conditions and the non-native disease white pine blister rust. Seed sources with durable genetic resistance to WPBR and the adaptive traits needed to survive warmer, drier conditions in the future are essential to sustain this species on the landscape. Seedlings from across the range of SWWP were tested for resistance to WPBR at Dorena Genetic Resource Center in Oregon. Major gene and quantitative resistance were documented. Data were also used to estimate resistance levels and frequency of resistance across the range of SWWP. Scion from parent trees identified as resistant is collected and grafted into John T. Harrington Forestry Research Center in Mora, NM. Seed will be collected from these orchards as well as resistant parent trees to be planted on the landscape. In addition, progeny identified as resistant at Dorena Genetic Resource Center will be grafted into a clone bank orchard at Tyrell Seed Orchard in Oregon to maintain these genetics. We have also established two common garden field trials in the Southwest to validate resistance results, monitor the long-term durability of resistance, and assess adaptive traits. This work provides information critical to identifying seed sources for future planting. Current challenges include funding ongoing activities (testing, outplanting and monitoring previous plantings), variable and unpredictable planting conditions, and mortality of original parent trees. This interdisciplinary, collaborative project includes international, academic, federal, and tribal partners. We will present details of the project, early results, and applications for management and restoration.

Conifer Regeneration Hinders Digital Estimation of Understory Plant Cover in Post-fire Whitebark Pine Communities

by Brandi E. Wheeler | Andrew J. Andrade | Elizabeth R. Pansing | Diana F. Tomback | University of Colorado Denver | University of Colorado Denver | University of Colorado Denver | University of Colorado Denver

Keywords: 1988 Yellowstone fires, digital image analysis, seral vegetation, subalpine forest, understory plant cover, visual estimation

Monitoring post-fire recovery of whitebark pine communities during climate change will enable us to determine the resilience of these communities. Evaluation of post-fire forest community recovery usually requires estimates of understory (non-conifer) plant cover. Photographic digital image analysis (DIA) is a commonly used method for estimating plant cover. However, DIA efficacy in multi-strata, post-fire plant communities may be reduced where visual obstructions (tree regeneration, coarse downed wood, and shadows) may conceal plant cover in digital imagery. We estimated recent understory plant cover in seral whitebark pine communities using permanent plots established at two study areas following the 1988 Yellowstone fires. Our goals were to 1) determine differences in visual obstructions between study areas in our digital imagery; 2) compare plant cover estimates derived from DIA, visual plot-level (20 m_2) estimation (PLE), and quadrat-level (1 m_2) estimation (QLE); and 3) determine relationships between estimated plant cover and visual obstructions measured *in situ*. DIA confirmed significant differences (odds ratio = 8.34) in percent conifer pixels between study areas. At the study area with fewer conifer pixels, DIA estimated, on average, 9% (95% CI = 3 – 14%) and 16% (95% CI = 10 – 21%) more plant cover than PLE or QLE, respectively, and had similar variability. At the study area with more conifer pixels, DIA estimated less plant cover than PLE or QLE by 28% (95% CI = 24 – 32%) and 22% (95% CI = 18 – 26%), respectively, and was more variable. Furthermore, subcanopy conifer regeneration was negatively associated with plant cover estimated by DIA but showed no relationship with visual estimates. We conclude that conifer regeneration hindered the detection of understory plant cover by DIA. We recommend visual estimation of plant cover in whitebark pine communities with multi-strata vegetation, although digital estimation may be advantageous early in succession.

Registrants

LeAnn	Abell		labell@blm.gov
Ilana	Abrahamson	US Forest Service, RMRS	ilana.l.abrahamson@usda.gov
Franklin	Alongi	Montana State University	franklin.alongi@gmail.com
Hannah	Alverson	Bureau of Land Management	halverson@blm.gov
Maurice	Amee		maurice.amee@uconn.edu
Elizabeth	Anderson	Waterton Biosphere Reserve	eanderson@watertonbiosphere.com
Andres	Andrade	University of Colorado Denver	andrew.andrade@ucdenver.edu
Vick	Applegate		k7vkmt@gmail.com
Steve	Arno		sfarno@msn.com
Charlie	Bacon		cbacon@usgs.gov
Gretchen M.	Baker	Great Basin National Park	gretchen_baker@nps.gov
Cliff	Bara		cliff.bara@whitebarkfound.org
Barbara	Bentz	USDA Forest Service, Rocky Mountain Research Station	barbara.bentz@usda.gov
Akriti	Bhattarai	University of Connecticut	akriti.bhattarai@uconn.edu
Joseph D.	Birch	Michigan State University	jcooper@ualberta.ca
James T.	Blodgett	USDA Forest Service	james.blodgett@usda.gov
Erika	Blomdahl	Utah State University	erika.blomdahl@usu.edu
John	Boetsch	National Park Service	john_boetsch@nps.gov
Enrico	Bonello	The Ohio State University	bonello.2@osu.edu
Andrew	Bower	USDA Forest Service	andrew.bower@usda.gov
Jennifer	Brickey		jennybe2@hotmail.com
Emily	Brodie	UC Davis Graduate Group in Ecology	egbrodie@ucdavis.edu
Harrison	Brookes		harrison.brookes@wyo.gov
Tom	Buckley		tnbuckley@ucdavis.edu
Karl	Buermeyer		karl.buermeyer@usda.gov
Lorinda	Bullington	University of Montana/ MPG Ranch	lbullington@mpgranch.com
Kelly	Burns	USDA Forest Service, R2, FHP	kelly.burns@usda.gov
Rich	Byron		rbyron@blm.gov
Blaire	Caldwell		blaire.caldwell@ucdenver.edu
Hilary	Cameron		hilary.cameron@pc.gc.ca
Amy	Campbell		amy.campbell2@usda.gov
Faith	Campbell		phytodoer@aol.com
Alina	Cansler	University of Washington	acansler@uw.edu
Buddie	Carroll		buddie.carroll@usda.gov
Charles V.	Cartwright	BC Ministry of Forests	charlie.cartwright@gov.bc.ca
Aramati (Anne Marie)	Casper	Colorado State University	aramaticasper@gmail.com
Natasha	Chadwell	Montana Technological University - Ecological Restoration	nchadwell@mtech.edu

Kristen	Chadwick		kristen.chadwick@usda.gov
Dana	Charlton		dana_charlton@tcenergy.com
Alana J.	Clason	Bulkley Valley Research Centre	ajclason@gmail.com
Christy	Cleaver	USDA Forest Service, Forest Health Protection	christy.cleaver@usda.gov
Jonathon	Colby		jonathon.colby@usda.gov
Philip	Collier		philip.collier@usda.gov
Owen	Comes		comesowen@gmail.com
Catherine	Conroy		cconroy@ktunaxa.org
Jennifer	Costich-Thompson		jennifer.costich-thompson@usda.gov
Cathy	Cripps	Montana state University	ccripps@montana.edu
Connor	Crouch		connor.crouch@nau.edu
Cory	Davis		cory.davis@umontana.edu
Liz	Davy	Forest Service	liz.davy@whitebarkfound.org
Ryan	Davy	Plumas NF	ryan.davy@usda.gov
Denyse	Dawe	Canadian Forest Service	denyse.dawe@canada.ca
Amanda	De La Torre		amanda.de-la-torre@nau.edu
Cindy	De Leon		cindy.battleborn@gmail.com
Chantelle	DeLay		chantelle.delay@usda.gov
Gregg	DeNitto	USDA Forest Service, R-1 FHP	gregg.denitto@gmail.com
Justin	DeRose	Utah State University	justin.derose@usu.edu
Kristina	DeYoung	USDA Forest Service	kristina.deyoung@usda.gov
Kimberly	Dohms	Canadian Wildlife Service	kimberly.dohms@ec.gc.ca
Bryan	Donner	US Forest Service	donnermt@yahoo.com
Joan C.	Dudney	University of California, Santa Barbara	jdudney@berkeley.edu
Justina	Dumont		justina.dumont@usda.gov
Tara	Durboraw	University of Colorado Denver	tara.durboraw@ucdenver.edu
Michael	Durglo		michael.durglo@cskt.org
Ashley	Durham		adurham@blm.gov
Cynthia	Dusel-Bacon		cdusel@icloud.com
Sherri	Elwell		sherri.elwell@gov.bc.ca
Blake	Engelhardt	USFS Region 5/Inyo NF	blake.engelhardt@usda.gov
Sebastian	Espinosa Novoa	Northern Arizona University	sje78@nau.edu
Eli	Estey		eestey@uidaho.edu
Beth	Fallon	National Park Service, Mount Rainier NP	beth_fallon@nps.gov
Brennan	Ferguson		brennan.ferguson@usda.gov
Allison	Fisher		allison.fisher@pc.gc.ca
Laurie	Fisher		laurie.fisher@usda.gov
Sarah J.	Flanary	RMRS Fire Sciences Laboratory	sarah.j.flanary@usda.gov
Robbie	Flower		robbie.flowers@usda.gov
John	Fothergill		jfothergill@blm.gov
Shelagh	Fox		shelagh.r.fox@usda.gov
Katy	Fraser	B.C. Ministry of Forests, Land & Natural Resources	katy.fraser@gov.bc.ca
Jeff	Galvin	NPS - MOJN	jeff_galvin@nps.gov
Amy	Gannon		agannon@mt.gov
Robin	Garwood		robin.garwood@usda.gov

Mike	Giesey	Whitebark Pine Ecosystem Foundation	mike.giesey@whitebarkfound.org
Sara A.	Goeking	US Forest Service, Rocky Mtn Research Station	sara.goeking@usda.gov
Betsy	Goodrich		anne.goodrich@usda.gov
Jeremy	Greenberg		jeremy.greenberg@pc.gc.ca
Emily	Guiberson	Bureau of Land Management	eguibers@blm.gov
Jennifer	Guse	National Park Service	jennifer_guse@nps.gov
Robin	Gutsell	Alberta Environment and Parks	robin.gutsell@gov.ab.ca
Sybille	Haeussler	Bulkley Valley Research Centre, UNBC	sybille.haeussler@unbc.ca
Anne	Halford	Bureau of Land Management	ahalford@blm.gov
Lacey	Hankin	University of Nevada, Reno	lacey.hankin@gmail.com
Andrew	Hansen		hansen@montana.edu
Chloe	Hansum	U.S. Fish and Wildlife Service	chloe_hansum@fws.gov
Susan	Harries		susan.harries@usda.gov
Cheri	Hartless	Bitterroot NF	cheri.hartless@usda.gov
Jill	Hautaniemi		jill.hautaniemi@mt.gov
Karen	Heeter	University of Idaho	kheeter@uidaho.edu
Scott	Heisler	NPS - Crater Lake	scottheisler@gmail.com
Amanda	Hendrix		amanda.hendrix@usda.gov
Emma	Heydenberk		emmaheydenberk@gmail.com
Justin	High		justin.high@usda.gov
Kevin	Hoekstra		kevin.hoekstra@gov.bc.ca
Trista	Hoffman		trista.hoffman@usda.gov
Lisa	Holsinger		lisa.holsinger@usda.gov
Sharon	Hood	USDA Forest ServicE Fire Sciences Laboratory	sharon.hood@usda.gov
Jen	Hooke	Crater Lake National Park	jen_hooke@nps.gov
Erin	Hooten	Bridger-Teton NF	erin.hooten@usda.gov
Beth	Horn		khcritters@aol.com
Brian	Howell	USFS, Forest Health Protection, Rocky Mountain Region	brian.howell@usda.gov
Sean	Hoy-Skubik	Montana State University	s.hoyskubik@gmail.com
Nicole	Hupp	BLM	nhupp@blm.gov
Tony	Incashola Jr		tony.incashola.jr@cskt.org
Francis	Iredale	Ministry of FLNROD	francis.iredale@gov.bc.ca
Katie	Ireland	Montana Department of Natural Resources	kathryn.ireland@mt.gov
Elizabeth	Jamison	National Park Service	elizabethjamison5@gmail.com
Cathie	Jean		terracasa45@gmail.com
Melissa	Jenkins	US Forest Service	melissa.m.jenkins@usda.gov
Chris	Jensen		chris.e.jensen@usda.gov
Jeremy	Johnson	Prescott College	jeremy.johnson@prescott.edu
Mike	Johnson		jay.m.johnson@usda.gov
Stephanie	Jouvet		stephanie.jouvet@gmail.com
Erik S.	Jules	Humboldt State University	erik.jules@humboldt.edu
Ellen	Jungck	USFS R1/R4	ellen.jungck@usda.gov
April	Kaiser	University of Idaho	akaiser@uidaho.edu
Jane	Kapler Smith		mtjanck@yahoo.com
Michael Edward	Kauffmann		michaelekauffmann@gmail.com

Robert	Keane	USFS, retired	bob.keane@whitebarkfound.org
Holly	Kearns		holly.kearns@usda.gov
Angelia	Kegley		angelia.kegley@usda.gov
Summer	Kemp-Jennings		summer_kemp-jennings@nps.gov
Kate	Kendall	Ursine Ecological	kkendall2382@gmail.com
Jeremy	Keys		jeremy.keys@usda.gov
Nickolas	Kichas		nkichas@gmail.com
Tom	Kimmerer		tom@kimmerer.com
Brian	Kittler		bkittler@americanforests.org
Vladimir	Kovalenko	University of Montana, National Park Service	vladimir.kovalenko@umontana.edu
Jodie	Krakowski		jodie_krakowski@hotmail.com
Donna	Kridelbaugh		donna.kridelbaugh@gmail.com
Dawn	LaFleur		dawn_lafleur@nps.gov
Gia	Landis		gkl37@nau.edu
Andrew	Latimer	UC Davis	amlatimer@ucdavis.edu
Jade	Lavoie		jadelavoie19@gmail.com
Rebecca	Lawrence	Glacier National Park	rebecca_lawrence@nps.gov
Kristin	Legg	National Park Service-Inventory & Monitoring	kristin_legg@nps.gov
Signe	Leirfallom		signe.leirfallom@umontana.edu
Adrian	Leslie		adrian.leslie@natureconservancy.ca
Jordan	Lestina		jordan.lestina@umontana.edu
Eric	Liner	Cornell	eliner@cornell.edu
Jun-Jun	Liu	Canadian Forest Service	jun-jun.liu@nrcan-rncan.gc.ca
Blakey	Lockman		irene.lockman@usda.gov
Laura	Lowrey		laura.lowrey@usda.gov
Matthew	Lufholm		mlufholm@mt.gov
James	Lutes	U.S. Forest Service	james.r.lutes@usda.gov
Nicole	Macy		nicole.macy@usda.gov
Colin	Maher	University of Alaska	cmbristlecone@gmail.com
Rob	Mangold	Whitebark Pine Ecosystem Foundation	rob.mangold@whitebarkfound.org
Enzo	Martelli	Restoration Ecology Lab, University of Montana	enzo.martellimoya@umontana.edu
Meghan	McFadden	Ulrich Lab	mcfaddml@whitman.edu
Anna	McIndoe		anna.mcindoe@gov.bc.ca
Shawn	McKinney	USFS RMRS	shawntmckinney@gmail.com
Thomas	McLaren	University of Colorado Denver	thomas.mclaren@ucdenver.edu
Charlie	McLellan		charlie.mclellan@pc.gc.ca
Michael	McWilliams		michael.mcwilliams@usda.gov
Connie	Mehmel	US Forest Service Forest Health Protection (retired)	cjmehmel@gmail.com
Sabine	Mellmann-Brown	USDA Forest Service, Region 1	sabine.mellmann-brown@usda.gov
Elliott	Meyer		elliott.meyer@usda.gov
Marc	Meyer	USDA Forest Service, Region 5 Ecology Program	marc.meyer@usda.gov
Elizabeth	Milano	USFS	ermilano@gmail.com
Amanda	Milburn		amanda.milburn@usda.gov
Hailey	Mizera	UNBC, BSc Wildlife & Fisheries	haileymizera@gmail.com
Emily	Monosson		emonosson@gmail.com

Brenna	Montagne	Shasta-Trinity National Forest	brenna.montagne@usda.gov
Annabelle	Monti	USDA Forest Service	annabelle.monti@usda.gov
Randy	Moody	Whitebark Pine Ecosystem Foundation of Canada	whitebarkrandy@gmail.com
Natalie	Morgan		natalie.a.morgan@usda.gov
Tanya E.	Murphy	Helena-Lewis and Clark National Forest	tanya.murphy@usda.gov
Michael	Murray	BC Ministry of Forests	michael.murray@gov.bc.ca
David	Neale		david.neale@whitebarkfound.org
Cara R.	Nelson	University of Montana	cara.nelson@umontana.edu
Nick	Nelson		nickpdx@gmail.com
Jonathan	Nesmith	US Forest Service PNW Research Station	jonathan.beals-nesmith@usda.gov
Maria	Newcomb	USFS Forest Health Protection	maria.newcomb2@usda.gov
Karen	Newlon		karen_newlon@fws.gov
Melissa	Nicolli	National Park Service Inventory and Monitoring Division	nicollim@oregonstate.edu
Russ	Oakes		russell.oakes@usda.gov
Brent	Oblinger		brent.oblinger@usda.gov
Kari	O'Leary		kari.oleary@usda.gov
Emma	Olney		emma_olney@yakama.com
Elizabeth	Pansing	American Forests	epansing@americanforests.org
Debbie	Paul		dkpaul@blm.gov
Lulu	Peach	Washington State University	lrp88@nau.edu
Monica	Perez-Watkins		mperezwatkins@nationalforests.org
Judy	Perkins		jlperkins@blm.gov
ShiNaasha	Pete	Confederated Salish and Kootenai Tribal Forestry	shinaasha.pete@cskt.org
Vernon S.	Peters	The King's University	vern.peters@kingsu.ca
Dave	Peterson		dave.peterson@usda.gov
Marta Irene	Pineda		irene@thegreenpine.com
Louis	Price		louis.price@gov.ab.ca
Diane	Probasco		diane.probasco@usda.gov
Michael	Pruss		michael.pruss@usda.gov
Teo	Rautu		teodora_rautu@hotmail.com
Chris	Ray	The Institute for Bird Populations	cray@colorado.edu
Charlotte	Reed	USDA Forest Service - Fire Sciences Lab	charlotte.reed@usda.gov
Michael	Reichenberg		michael.reichenberg@usda.gov
Iain R.	Reid	Parks Canada Agency	iain.reid@canada.ca
Jason	Reinhardt		jason.reinhardt@usda.gov
Austin	Rempel		arempel@americanforests.org
Katherine	Renwick		katherine.renwick@usda.gov
Emily	Rhoades	USFS Coeur d Alene Nursery	emily.rhoades@usda.gov
Chad	Riffle		chad.riffle@usda.gov
Rebecca	Rising	North Cascades National Park	rebecca_rising@nps.gov
Maegen	Rochner	University of Louisville	maegen.rochner@louisville.edu
Steve	Rust		srust@naturescap.com
The	Romero	Universidad Michoacana de San NicolÃ¡s de Hidalgo	csaenzromero@gmail.com
Jessie	Salix		jessica.salix@usda.gov
Rinki	Sarkar	Independent Researcher	rinkisarkar@gmail.com

Peri	Sasnett	Glacier National Park	peri_sasnett@nps.gov
Taza	Schaming	Northern Rockies Conservation Cooperative	tazaschaming@gmail.com
Anna W.	Schoettle	USDA FS - RMRS	anna.schoettle@usda.gov
Bruce	Schuelke		bruce.schuelke@usda.gov
Glenda	Scott		events@whitebarkfound.org
Glenda	Scott	Whitebark Pine Ecosystem Foundation	glenda.scott@whitebarkfound.org
Casey	Seaman	USDA Forest Service, R1, Nez Perce - Clearwater NF	casey.seaman@usda.gov
Julee	Shamhart	Whitebark Pine Ecosystem Foundation	julee.shamhart@whitebarkfound.org
Erin	Shanahan	NPS-Greater Yellowstone Inventory and Monitoring Network	erin_shanahan@nps.gov
Brenda	Shepherd	Parks Canada	brenda.shepherd@canada.ca
Cindy	Simmons		cindy.simmons@usda.gov
Rachel	Simons		rachel.simons@usda.gov
Laurel	Sindewald	University of Colorado Denver	laurel.sindewald@ucdenver.edu
Robert	Sissons		robert.sissons@pc.gc.ca
Mindy	Skinner		mindy.skinner@pc.gc.ca
Genevieve	Skora		genevieve_skora@fws.gov
Michele	Slaton	US Forest Service PSW Remote Sensing Lab	michele.slaton@usda.gov
Rebecca	Smith	Parks Canada	rebeccaceline.smith@pc.gc.ca
Sean	Smith		sean_b_smith@nps.gov
Stu	Smith		wagmorebarkless@protonmail.com
Brian	Smithers		brian.smithers@montana.edu
Richard A.	Sniezko	USDA Forest Service	richard.sniezko@usda.gov
Lisa	Solberg Schwab		lisa_solbergschwab@fws.gov
Ben	Solvesky	U.S. Fish and Wildlife Service	ben_solvesky@fws.gov
Hanno	Southam	Simard Lab, University of British Columbia, Vancouver	hannosoutham@gmail.com
James H.	Speer	Indiana State University	jim.speer@indstate.edu
Eric	Sprague	American Forests	esprague@americanforests.org
Natalie	Stafl	Parks Canada	natalie.stafl@gmail.com
Chris	Stalling		chris.stalling@usda.gov
Elisa	Stamm	USFS - R1 - IPNF	elisa.stamm@usda.gov
Cameron	Stauder		cameron.stauder@usda.gov
Cathy	Stewart	USFS retired	mtstew54@yahoo.com
Megan	Strom		megan.strom@usda.gov
Devin	Stucki	National Park Service	devin_stucki@nps.gov
Wes	Swaffar		wswaffar@americanforests.org
Lauren	Taracka	Central Washington University	tarackal@cwu.edu
Peter	Tarleton		peter.tarleton@pc.gc.ca
Robin	Taylor-Davenport		robin.taylordavenport@usda.gov
David	Thoma	National Park Service Inventory and Monitoring Program	dave_thoma@nps.gov
Diana F.	Tomback	University of Colorado Denver	diana.tomback@ucdenver.edu
Danielle	Ulrich	Montana State University - Ecology	danielle.ulrich@montana.edu
Marc	Valencia		marc.valencia@usda.gov

Phillip	van Mantgem		pvanmantgem@usgs.gov
Eric	VanderBeek		eric.a.vanderbeek@usda.gov
Joanne	Vinnedge	WPEF-Canada	vinnedgj@telus.net
Matthew O.	Voigt	Helena-Lewis and Clark National Forest	matthew.voigt@usda.gov
David	Walker		davidwalkermember@gmail.com
Kristen	Waring	Northern Arizona University	kristen.waring@nau.edu
Chloe	Wasteneys	Montana State University	chloewasteneys@montana.edu
T (Tad)	Weaver	Mont State Univ	tweaver@montana.edu
Brandi E.	Wheeler		bwheeler@nmsu.edu
Max	Wheeler		mackenzie.wheeler@usda.gov
Jessica	Wiedeman		jessica.wiedeman@usda.gov
Nicholas	Wilhelmi		nicholas.wilhelmi@usda.gov
Beth	Willhite		beth.willhite@usda.gov
Geoffrey	Williams		geoffrey.williams@usda.gov
Tyler J.	Williams	Community College of Denver	tyler.williams@ccd.edu
Brendan	Wilson	selkirk college	bwilson@selkirk.ca
Lisa	Winn	USFS Dorena Genetic Resource Center	lisa.winn@usda.gov
Beth	Wood		elizabeth.wood@usda.gov
Carolyn	Wright	Royal Roads University	carolyncwright27@gmail.com
Art	Zack		artzack@gmail.com
Solomon	Ziegert		solomonrexziegert@gmail.com

Made in the USA
Monee, IL
02 November 2026

67385502R00105